THE LANGUAGES OF RELIGION

This book analyses the power that religion wields upon the minds of individuals and communities and explores the predominance of language in the actual practice of religion. Through an investigation of the diverse forms of religious language available – oral traditions, sacred texts, evangelical prose, and national rhetoric used by 'faith-insiders' such as missionaries, priests, or religious leaders who play the communicator's role between the sacred and the secular – the chapters in the volume reveal the dependence of religion upon language, demonstrating how religion draws strength from a past that is embedded in narratives, infusing the 'sacred' language with political power.

The book combines broad theoretical and normative reflections in contexts of original, detailed and closely examined empirical case studies. Drawing upon resources across disciplines, the book will be of interest to scholars of religion and religious studies, linguistics, politics, cultural studies, history, sociology, and social anthropology.

Sipra Mukherjee is Professor in the Department of English, West Bengal State University, India. Her research interests are religion, caste, and power. Her publications include *Interrogating My Chandal Life: An Autobiography of a Dalit* (translation of Manoranjan Byapari's *Itibritte Chandal Jeeban*, 2018), *Modern English Literature, 1890–1960* (2016), Special Issue on Religion and Language, *International Journal of the Sociology of Language* (edited, 2013), and *The Calcutta Mosaic: Minority Communities of Calcutta* (co-edited, 2009).

THE LANGUAGES OF RELIGION

Exploring the Politics of the Sacred

Edited by Sipra Mukherjee

Routledge
Taylor & Francis Group
LONDON AND NEW YORK

First published 2018
by Routledge
2 Park Square, Milton Park, Abingdon, Oxon OX14 4RN

and by Routledge
711 Third Avenue, New York, NY 10017

Routledge is an imprint of the Taylor & Francis Group, an informa business

British Library Cataloguing-in-Publication Data
A catalogue record for this book is available from the British Library

Library of Congress Cataloging-in-Publication Data
A catalog record has been requested for this book

ISBN: 978-0-415-31604-0 (hbk)
ISBN: 978-0-429-46592-5 (ebk)

Typeset in Galliard
by Apex CoVantage, LLC

CONTENTS

CONTRIBUTORS

Chad M. Bauman is Professor of Religion, and Chair of the Department of Philosophy, Religion, and Classics at Butler University, Indianapolis, Indiana, USA. He is currently researching conflict between Christians and adherents of other faiths in India and Sri Lanka. He wrote *Christian Identity and Dalit Religion in Hindu India, 1868–1947* (2008) and *Pentecostals, Proselytization, and Anti-Christian Violence in Contemporary India* (2015). He is also co-editor (with Richard Fox Young) of *Constructing Indian Christianities: Conversion, Culture, and Caste* (2014).

Nelly Bekus is Associate Research Fellow at University of Exeter, UK. She was a postdoctoral fellow at the Davis Center at Harvard University (2012–13) and has worked as an assistant professor at the University of Warsaw (2008–12). She wrote *Struggle over Identity: The Official and the Alternative "Belarusianness"* (2010), and (with Michal Wawrzonek and Mirella Korzeniewska-Wiszeniewska) *Orthodoxy Versus Post-Communism? Serbia, Belarus, Ukraine and the Russkiy Mir* (2016). Her research interests include state-building under socialism and post-socialism, the role of cityscape in the construction of post-socialist identity, and the religious and the ethno-linguistic landscape in post-Soviet nations.

Marta Dominguez Diaz is a Senior Lecturer in Islamic Studies at the University of St Gallen (Switzerland) and Senior Research Fellow at the Department of Religious Studies of the University of Vienna. She has previously held research and teaching posts at SOAS (London) and the Woolf Institute (Cambridge). Her research interests include cultural identities and minorities in North Africa, North-African Sufism, transnational Islam, Islam in Europe, comparative religion and Muslim-Jewish relations. She has published a number of academic articles, *Women in Sufism: Female Religiosities in a Transnational Order* (2014), and *Tunisia's Andalusians: The Cultural Identity of a North African Minority* (forthcoming).

Gilad Elbom is Senior Instructor in the School of Writing, Literature, and Film at Oregon State University, USA. His first novel, *Scream Queens of the Dead Sea* (2004), was translated into German, Russian, and Hebrew. His fiction and scholarship have appeared in *Arts & Letters, The American Journal of Semiotics, The Journal of Communication and Religion, Portland Review, Santa Monica Review*, and *ZYZZYVA*.

Charlotte Eubanks is Associate Professor of Comparative Literature, Japanese, and Asian Studies at Pennsylvania State University, USA, where she studies transnational Buddhism, word/image relations, and the material culture of books, with a focus on Japanese literature from the medieval period to the present. She is the author of numerous articles and the monograph *Miracles of Book and Body: Buddhist Textual Culture and Medieval Japan* (2011), and she is associate editor at the journal *Verge: Studies in Global Asias.*

Jose Mario C. Francisco is Professor of Philosophical and Systematic Theology at Loyola School of Theology, Ateneo de Manila University, Philippines. His research into Asian contexts of religion places theological themes in their historical contexts and explores their reception in different linguistic and cultural contexts. Fundamental to his work in these two areas is the paradigm drawn from contemporary translation theory – that Christianity is translated as it moves from one context to another.

Epsita Haldar is Assistant Professor at the Department of Comparative Literature, Jadavpur University, Kolkata, India. She is Charles Wallace India Trust visiting fellow at the School of Oriental and African Studies, University of London (2018). She has been working on the Muharram traditions in West Bengal, a part of which has been done with the Art Research and Documentation Grant of India Foundation for the Arts, Bangalore. She is a recipient of the Charles Wallace Trust Short Term Fellowship and Sarai-CSDS Social Media Fellowship. She is interested in the interface between Muslim popular piety, visual culture, and new media in South Asia.

Amanda J. Haste is adjunct faculty at Aix-Marseille University, France, where she teaches courses in translation and in English as a musicological language. Her doctoral thesis (Bristol, 2009) was on the role of music in twenty-first-century Anglican monasticism, and she publishes regularly on monastic music, music and identity, and translation issues. She co-edited (with James Block) *Constructing Identity in an Age of Globalization* (2015) and contributed six articles to the *Encyclopedia of Women in World Religions*, ed. Susan De Gaia (2018). She is a member of the Chartered Institute of Linguists, and President of the National Coalition of Independent Scholars.

Andreas Koechert is Professor of Cultural Studies at the University of Quintana Roo, Chetumal, Mexico. His research interests include ethnicity, religion, social change, and sociolinguistics of indigenous and mestizo societies in Guatemala, Mexico, and Peru. His long-term fieldwork during the internal armed conflict in Guatemala gave access to the ritual knowledge of Kaqchikel spirituality. His forthcoming book is on Kaqchikel prayers of life and death.

Kerith Miller is currently a PhD candidate at the University of Arizona, USA, working on a dual degree in linguistic anthropology and Middle Eastern studies. She has obtained dissertation research fellowships from the German Academic Exchange Service (DAAD), Orient-Institut Istanbul, and the University of Arizona School of Middle Eastern and North African Studies from August 2015 to April 2017, to examine the use of both Syriac and Surayt (Turyoyo) by the Syrian Orthodox diaspora communities in Istanbul and Berlin, focusing on the revitalization of Surayt. She has previously published in the journal *Perspectives on Arabic Linguistics* (2014) about Uzbekistan Arabic.

Lionel Obadia is Professor of the Anthropology of Religion at the University of Lyon, France. He has been studying Buddhism (in the West and in Asia), shamanism and witchcraft (in Asia), Jewish Messianic movements, globalization/modernity and religion, and new paradigms in the study of religion and belief. He is the author of ten books and a hundred papers, including "Buddhism, Globalization and Modernization" (2014), "Globalization and the Sociology of Religion" (2010), and "The Economics of Religion" (2011).

Barbara Pfeiler is Senior Researcher at Centro Peninsular en Humanidades y en Ciencias Sociales – Universidad Nacional Autónoma de México, Mérida, Mexico. She received her PhD in Romance Philology from the University of Vienna and the Habilitation in Mesoamerican Studies from the University of Hamburg. Her interest in ritual language goes back to her participation in interdisciplinary projects on Maya spirituality in Mexico and Guatemala.

INTRODUCTION

The languages of religion

Sipra Mukherjee

The contemporary age is possibly the first in the history of civilization to witness a general recognition of religion as a political subject. Although people have, since ancient times, been aware of religion's potential as a political force, the perspicacity and astuteness to use that knowledge was always limited to a few, usually powerful individuals. These were often political or religious leaders who could clothe themselves in religious or secular garb with equal ease, employing the languages of religion to speak to the many who were less observant of the connections between religion and politics.

A number of events up through the twentieth century – most of them unfortunate – have, however, transformed this scenario. Religion and politics are now commonly recognized as overlapping spheres by a wide population comprising the powerful and the powerless, the religious and the secular. Leaders and scholars of religion who, till the early twentieth century, debated and argued over ecclesiastical aspects of faiths now spend as much time discussing the political implications of religious language and practices. Whereas traditional forms of scholarship assumed that religious customs, language, and texts were "politically neutral rather than expressions (or claims or disguises) of relations of power" (Green and Chatterjee 2008: 1), the modern age has witnessed many explicit claims to political power articulated through religious language. These claims – sometimes subtle and masked, other times unvarnished and naked – have been made both from within the traditional echelons of power (the institutions of state and religion) and from the margins. With modern technology perforating the once-impregnable walls that lined the corridors of power, the late twentieth and early twenty-first centuries have heard many 'other' voices whose language blends together religion and politics.

The area of political religion, or politicized religion, received little critical attention until the 1980s, despite the religious basis of Hitler's Fascism and the Armenian genocide. It is significant that the term 'religious terrorism' is fairly recent, coined in the context of Islamic terrorist attacks on the First World. Religion and nation were, prior to this, seen as allied arenas, sharing the same territory, and the threat of terrorism was perceived as a threat from outside this territory. 'Terrorism', a word that has insidiously worked its way into our everyday vocabulary (Hoffman 2006), was thus a political or secular phenomenon and a threat from 'outside' the nation, from those who were foreign in both religion and nationality. Since the 1980s, however, the definition of 'terrorism' has been largely rewritten, with religious and national identity not necessarily implying one single, simple group. As a consequence, 'terrorism' now springs from domestic as well as foreign shores. Using the languages of religion, radical groups, sects, and cults have targeted others of their own nation and religious community as well as those of other nations and religious communities. Given this reality, the languages of religion demand study and analysis by scholars of various domains. Sociologists, historians, anthropologists, linguists, political scientists, psychoanalysts, and culture studies experts have lent their expertise to the study and analysis of the languages of religion.

In almost every recent scholarly work on the narratives of religion, scholars have argued that the subject has more to do with power than with religion per se. Green and Searle-Chatterjee (2008) showed how the genealogy of religious studies has marked 'religion' as a distinct territory which has remained largely blind to the political implications of these systems of thought. The study of religion has tended to assign key positions to the sacred texts and 'insiders' of a faith in order to understand a religion. "By resorting to texts and persons (who were in fact engaged in all manner of scrambles for power), it was thought that an 'objective' description could be given of the proper constituent elements of any particular 'religion', which could then be identified by its key texts, experts, and institutions and kept distinct from the constituents of other 'religions'" (1–2). Such scholarly formations, Green and Searle-Chatterjee argue, are not only intrinsically ideological, but often blatantly ahistorical. Viewing religious texts, written or oral, as 'sources', passive media that shed light on the inner world of 'belief' or the outer world of 'events', ignores the agency and power that these texts possess to make 'moves' in the world, and so to reshape it (2008: 2).

The 'audacity' to study religion received impetus with Marx's 1843 statement that religion was the opium of the people. Though usually understood to be a criticism of religion, and in consequence a criticism, albeit a sympathetic one, of opium addicts, these words really point to the rather significant role played by religion in human lives. They also point to the role that language plays in concretizing the abstraction of religion. The original lines from Marx read:

> Religion is the general theory of this world, its encyclopaedic compendium, its logic in popular form, its spiritual *point d'honneur*, its enthusiasm, its moral sanction, its solemn complement, and its universal basis of consolation and justification. . . . *Religious* suffering is, at one and the same time, the *expression* of real suffering and a *protest* against real suffering. Religion is the sigh of the oppressed creature, the heart of a heartless world, and the soul of soulless conditions. It is the *opium* of the people.
>
> (1844, italics in original)

Religion articulates, largely through language, the 'expressions' of 'protest' and the 'sighs' of 'suffering' that make this "logic in popular form" accessible to a people. Communities encounter religion through various genres of language: ritual-sanctified words or consecrated narratives such as parables, hymns, chants, and myths. With its powerful status as the major and most flexible instrument of communication, language is indispensable to the expression of religion and religiosity. Existing largely in its languages, the power of religion is enhanced by the shifting and fluid nature of language. Indeed, the apparent materiality of language, with its aural sounds and visual images, belies the lack of fixity that characterizes words – a stability of meaning that is promised but seldom delivered. Language may, perhaps, be described, in Lacanian terms, as an imaginary pursuit of illusory certitudes. This lack of certitude, and indeed enduring ambiguity, inherent in words functions as an apt vehicle for religion, a domain that draws upon the unknowable and the impenetrable for its strength.

While some of the chapters of this book reveal the influence which religion and language have had on each other at various times and in various places, others speak of the potential of politics that exists between the two. Such politics is made possible by the realities that govern the processes of institutionalization of religion in every community and nation. All communities invest in the disciplining of minds and in the making of memories. It would be naïve to believe that such investments – ideological, institutional, and political – would leave unaffected "the texture of produced historical knowledge" (Kar, n.p.). With the onus of fashioning the imagined community repeatedly placed on religion, the trajectory of a nation could hardly remain untouched by the high priests of faith. In modern multi-religious communities too, where the forces of modernity and transnational realities have propelled humanity towards an inclusive cultural identity, the imagined community has been articulated through language borrowed from religion, revealing religion's tacit but significant role in the making of our social imaginaries. The present collection of essays will approach these issues using case studies that reveal the negotiations – sometimes subtle, other times blatant – between languages and religions.

Most religions begin with the unconditional belief that the origin of religion is divine and conclude with the conviction that it is hence true and unquestionable. Any secular analysis of religion subverts this belief by beginning with the argument that man predates religion. Modern scholarship on religion contends that man made religion; religion did not make man. To extend the argument, therefore, religion serves man, not the other way around. Yet, despite the rationalist naysayers, despite the progress of science and technology, and indeed despite the many gaffes by religions' own prophets – the innumerable apocalyptic denunciations followed by the anti-climactic continuation of a world that endured – religion has remained a powerful force dominating the minds of men and women. Scholars have historicized it, contextualized it, and analysed it threadbare, yet they have been unable to denude it of its mystical sheen and its magnetic power.

What makes religion so resistant? Where lies the power of religion? In the post-Darwinian world, the idea of modernity – with its concomitant ideas of science, technology, secularism, individualism, and democracy – has been seen as engendering a space inhospitable to religion. Increasingly, however, religion has appropriated these ideas, moving with them, not against them as was imagined till the mid-twentieth century.

To approach 'religion' as an innocent term, one of the many foundational terms that categorize our spheres of understanding – like 'psychology', 'political science', or 'economics' – would be misleading. Like the word 'science' – a word that at one time meant 'knowledge' but has now come to mean only one kind of knowledge, hence privileging a particular area of learning – the word 'religion' is more than a mere indicator of a certain kind of content. It is a word imbued with pre-determined values that privilege and announce its subject. Popular etymology amongst the later ancients – Servius, Lactantius, and Augustine – connects the word 'religion' with *religare*, "to bind fast". St Augustine, whose writings shaped the Latin Church and Western philosophy, also made this derivation, in accordance with the Christian writer Lactantius, considering the root *ligare* to be at the root of 'religion' (Hoyt, 126.) Cicero, though, derived the word 'religion' from *re-ligere*, "to gather up again, to take up, to consider, to ponder – opposed to nee-legere, to neglect." Max Muller wrote, "I belief myself that Cicero's etymology is the right one; but if religion meant originally attention, regard, reverence, it is quite clear that it did not continue long to retain that simple meaning" (11–12). The English meaning of the word – particular systems of faith which entail reverence and piety, with wide-ranging associations in the social, political, and spiritual domains – came much later and is recorded from around 1300 CE.

The history embedded in the term points to old and intricately laid roots in the minds of human beings and in their histories of bonding with each other. Religions have been instrumental in constructing identities and have been implicated in wide-ranging formative processes, expressing not only individual personhoods but also collective alliances. Within such trajectories of individual/collective formations, language is the crucial lens through which identities are perceived, experienced, and articulated. The discursive affiliations and underlying epistemologies of words and phrases anchor the metanarrative of the religious identity constructed for an individual or a collective.

With these questions in mind, this volume brings together case studies that explore the making of religions' languages. Drawing upon expertise in linguistics, sociology, culture studies, and the social sciences, the volume sets broad theoretical and normative reflections in contexts of closely examined case studies. Taken together, the chapters argue that a scholarly approach to religion requires an understanding of the intimate connection that exists between the language that a religion uses and the power that that language wields over the minds of a community. The case studies investigate the diverse forms of religious language – oral traditions, sacred texts, evangelical prose, and nationalist rhetoric used by 'faith-insiders' such as missionaries, priests, and religious leaders who liaise between the sacred and the secular. The essays reveal the dependence of religion upon language, demonstrating how religion draws strength from a past that is embedded in narratives, infusing the 'sacred' language with political power. Using historical and/or ethnographic content of their subject areas, the authors employ a variety of theories and methods, bringing their differently tuned expertise to the subjects they discuss. They cover subjects that go beyond the usual religious narratives such as parables or ecclesiastical doctrines. While the chapters deal with specific instances of how religion has used language at various times and in various spaces of history, the broader aim of the book is to indicate the potential of politics that exists within them. Religion and language are both bridges constructed by humanity to reach objects that are notoriously inaccessible. Yet the discourses of both have been mired in, and indeed have drawn sustenance from, worldly politics of the material and the visible. Such politics, encapsulated in the shifting relations involving authority and power, is made possible in the ideological world each inhabits, through words and rituals that seldom have unambiguous meanings; where connotations and interpretations are slippery and shifting; and where their very ambiguity and imprecision lend power to the words and rituals of the languages of religion.

This fluidity of connotation possible in language, and the broad scope this may make possible for religion, is most easily visible in contexts of religious conversion, and six of the eleven essays explore either contexts of conversion or contexts of the movement of a religion's followers to another country. Through explorations of the intricacies of language, these case studies reveal that traditions of the word function in ways that are both inherited and invented. The cultural legacy that a word carries is embedded with images and memories that are a part of the social imaginary of the community. While these contribute to the inherited tradition of that word, and constitute its chief force, they sometimes conflict with the new cultures within which the word is placed, necessitating negotiations and inventions of new traditions that harmonize with the new context. The essays have been grouped accordingly. The first group deals with studies that highlight inherited language traditions that for centuries have empowered and enabled religions, and the next group deals with studies that reveal processes, usually over years, that have invented language traditions in accordance with communities' needs. This neat division of the essays into categories is, however, anything but. The inherited and invented traditions overflow onto each other, with the 'new' invented and the 'old' inherited aligned diachronically, with the new appearing always, as Benedict Anderson wrote of a related sphere, "to invoke an ambiguous blessing" from the old (1991: 187). This is borne out by some of the essays in the second section that deal with the twists and turns in the trajectory of traditions, as newer 'invented' elements are infused into inherited traditions to render them acceptable in modern or changing contexts. Similarly, the first section, on inherited traditions, sometimes bears out Eric Hobsbawm's argument that those traditions "which appear or claim to be old are often quite recent in origin and sometimes invented" (1983: 1). The third group of essays deals with studies that look at contexts of politicization of religious languages, wherein extra-linguistic factors such as cross-cultural tension or a nation's need to construct legitimacy and find validation have been met by the languages of religion.

Three of the four essays in the first section, dealing with inherited language traditions are, in apparent contradiction to the theme, on religious conversion. This is because the existence of inherited traditions in the continued life of a religion amongst its followers is an expected and unexciting phenomenon. It is the concept of inherited traditions persisting in narratives of a change of religion that piques the scholar's attention. The focus in this section is twofold: first, on traditions that are proclaimed as new, or as inherited, but which a more in-depth study may reveal as only partially so;

and second, on the intricate manoeuvres of inherited traditions to validate the new faith of the converted.

Syncretic forms of worship and the historical peculiarities of sacred languages resulting from the spread of religion inform the first two essays in this section. The essays, one from the Far East and the other from the far West, are rigorous studies of sacred reading materials, prayers, vocal utterances, sounds, and political contexts of faith. The first is by Charlotte Eubanks, who deals with the language of Buddhism in China and Japan, which has remained one of the great curiosities of East Asian Buddhism. Eubanks reveals how sacred languages do not necessarily wield power through the sense they make; rather, they work subliminally through sounds, often meaningless, whose associations we are only barely conscious of. Eubanks uses as her case study the reign of a Japanese emperor who was pivotal in transitioning the reading of the Buddhist sutras from an activity that monks engaged in as a spiritual, temple-based praxis to a cultural and artistic pursuit that could be used as cultural capital by imperial Japan. When, in the early centuries CE, Central Asian monks rendered the Buddhist scriptures into a host of languages, including Chinese, they chose to translate some terms while deciding to merely transliterate some others. The sutras thereafter remained till today, by and large, in the language of Classical Chinese, despite an awareness of both the presence of Sanskrit originals and the advantages of translation into local languages such as Japanese. Eubanks argues that despite, or because of, the emptying out of meaning of certain untranslated words, the aura around the reading of the sutras was heightened. The musical recitation of sutras in the Japanese court, which involved dwelling on, swelling out, breaking through, and melismatically extending the sound of each phoneme, rendered the reading more about performing a vocal text, about producing the sounds of sacred scripture, than about making grammatical sense.

The second essay, by Andreas Koechert and Barbara Pfeiler, shows how the peculiar plurality of the Catholic language in Central America makes it possible for Maya religiosity to survive amongst the Kaqchikels, an ethnolinguistic group, and may also make the community resistant to penetration by the Catholic faith and its verbal doctrine. The authors use anthropological and linguistic data to show how long-held worldviews and cultural behaviour may have a greater hold on the minds of communities than newly learned 'modern' faiths. Possibly as a result of the lack of priests in rural parts of Guatemala, which caused the responsibility of the Mission to fall on an organization called *cofradía*, the language of Kaqchikel Catholicism encourages intercultural semantics to achieve a 'common' understanding of the religion. Analysing specific words that show a mutual interaction between Catholicism and Kaqchikel Maya spirituality, in form as well as in content, the authors indicate that the Kaqchikel ritual language can be understood as an expression of both the Christian and the Maya faith, revealing an intertwining of religious behaviour, and the interpretation of many Catholic elements in accordance with Maya spirituality.

Chad Bauman's essay, which examines the dynamics of religious conversion amongst Dalit Satnamis in a colonial province of central India, studies the methods used by missionaries to show how the Satnamis' preconceived notions and insecurities were played upon by European missionaries to spread Christianity. Denied access to religious texts through Hinduism's tradition of discrimination, the Satnamis were painfully conscious of their general illiteracy and lack of a sacred text. The authority of written texts, as above that of oral traditions, and access to the Holy Book were therefore distinctive and deliberate aspects of the missionary epistemology that attracted this marginal community. In their attempts to make the new faith a part of the Satnami collective imaginary, missionaries attempted to forge connections between their faith and that of the Satnamis. This was done, amongst several ways, by attempting to portray Christianity as the culmination and fulfilment of the Satnami guru's message. Christianity was, therefore, projected not as 'the new true faith' but as a fulfilment of the promised future of the old one. Such a portrayal required selective editing and rewriting of Satnami history and the re-mythologization of the Satnami guru as John the Baptist. What emerges from this study is, therefore, a striking mélange of religious mythologies, wherein supposedly age-old tales and reverence for the word come together to endorse the authenticity of the new faith.

A similar story of Christianity transforming a predominantly oral native society by transposing its native script into the Roman alphabet and producing religious texts in native languages occurred in the Philippines. Unlike Bauman's study, which focuses on the content of the texts, Jose Mario C. Francisco's study focuses on the central role that linguistic aspects play in the construction of Filipino Christianity. Francisco argues that the extensive social presence of Christianity in Filipino society is due to the incorporation of the centuries-old Philippine oral narrative tradition of reading and performing texts. This permitted an indigenization of the faith, making possible the emergence of a Christianity rooted in native society. Filipino society is characterized by a traditional oral-aural culture in which, in Ricoeur's words, "the dialectic of event and meaning" in discourse remains tied to its particular interlocutors. The creation of religious texts in local languages enabled Filipino converts to approach the texts differently from literate followers of Christianity in the West but in accordance with their traditional narrative techniques which facilitate orality. With its antiphonal chanting that breaks down boundaries between narrator and listener, Filipino Christianity involves the congregation of kin and neighbours as chanters who take turns at being narrator/preacher and listener/preached-to, thus re-imagining the familiar social context within which earlier faiths were rooted.

The second section, dealing with invention of traditions, begins with an essay by Lionel Obadia that researches the liminal nature of such 'inventions' and depicts the intimate reciprocity between inheritance and invention of traditions, even as it showcases the conflicts that may be engendered between them. With its focus on the alterations and modifications that inherited traditions undergo to ensure their acceptance over ages and across cultures, this study traces the journey of Buddhist communities from Asia to the West. Obadia delves into the complex processes of translating religion from one culture to another as precepts and histories associated with words colour and appropriate ideas. He emphasizes the tension between idiomatic and cultural frameworks in the diffusion of Buddhist ideas as he analyses the role of language in the diffusion, adaptation, and transformation of Asian religious practices and institutions in the secular West. What happens to traditional themes, he asks, as the religion is interpreted from Tibetan, Chinese, or Japanese to English, French, or German? As Buddhist themes and ideas are reinterpreted through a series of specific repertoires, corresponding with mainstream Western cultural frameworks (scientism and humanism), prominent systems of knowledge (physics and psychology), or, more recently, ideologically oriented frameworks (feminism, postmodernism, and ecology), Western Buddhism becomes contested territory as it is laid claim to by the often-conflicting trends of modernism and traditionalism.

The second essay of this group, by Amanda Haste, takes this study a step further by introducing Ricoeur's concept of "narrative identity" and linking the construction of sacred languages with that of identity. Haste looks into the making of the Anglican linguistic identity with the Oxford Movement three centuries after the English king Henry VIII broke with Rome and created the Church of England. Choosing vernacular English as the ritual language of the new Church, King Henry VIII underlined the break with Catholicism by rejecting the Latin of the Catholic service books in favour of a tongue "clearly understanded of the people". In the 1840s, when the Oxford Movement led to a reconsideration of the Church of England's relationship with the Roman Catholic Church, there began for the monastic communities a problematic linguistic journey as they negotiated their Catholic roots while creating a recognizably Anglican identity. Haste follows the continuing search for a narrative containing "within one body both Protestant and Catholic elements" from the nineteenth into the twenty-first century, emphasizing how this continuing search has been a linguistic challenge for the Anglican community.

The next essay is a case study by Gilad Elbom of a modern effort to invent a cult that would serve to keep people away from nuclear waste repositories that will likely remain radioactive for more than ten thousand years. The United States Department of Energy's concern about these repositories led experts to conclude that the best system would be one that used elements of religion and myth to keep future generations away. Given the likelihood that these repositories will remain radioactive long after current methods of communication, including natural languages, are no longer understood, there is a need for a system that will remain effective over, so to say, eternity. The rationale behind the invented cult is a knowledge guarded by an elite group, akin to "an atomic priesthood", who will lead the rest of the population, "the uninitiated", into following a tradition of faith through remembrance and retelling. In semiotic terms, writes Elbom, a site whose deadliness cannot be named or specified "is a signified without a signifier", similar to another such signifier that indicates an absent signified: the Hebrew deity. Using this comparison as a springboard, Elbom analyses the language of a few key biblical narratives to indicate linguistic, literary, and semiotic modes through which the survival of the Jewish faith has been facilitated.

It has not always been possible to resolve the search for a narrative that validates both religious and national identities with such concern for uniformity, and Marta Dominguez Diaz's essay deals with the plurality that has emerged in the transnational Sufi Order Qādiriyya al-Būdshīshiyya. Originating in a Moroccan *wilāya*, in recent years the faith has spread across West Africa and Western Europe to Latin America, the United States, and Canada, accepting into its fold Arab populations, Moroccan labour migrants in Europe, European sons and grandsons of Muslims who see it as a way back to practising Islam, and European converts to Islam. Diaz argues that addressing such a varied audience has been made possible by a proactive proselytizing endeavour, and central to this enterprise has been the production of written materials in languages as varied as Arabic, French, Spanish, and English. It is significant that these publications are neither texts of doctrine or religious law nor translations of the same texts but mainly novels and members' autobiographies, with notable differences in content, style, and understanding of the faith, which construct a diverse tradition of faiths. Far from possessing the universalistic appeal that Islamic and Sufi parlance usually adopts, this Sufi Order's texts reveal that both the targeted reader and the content may be diverse across language groups. Diaz goes on to argue that despite several Būdshīshiyya subgroups emerging with markedly different conceptions of Islam, a communal religious identity is shared by followers, who believe they are, despite their different languages, sourcing their creeds from the same fount of faith.

The last section consists of three case studies set in contexts wherein the languages of religion have been impacted by extra-linguistic factors such as the breaking and making of nations, language policies, international tensions, modernization, the collective imaginaries of a community, and predefined notions about certain religions.

The first, by Nelly Bekus, is a study of Belarus after the fall of the USSR, when religion returned to public life after nearly seventy years of aggressive atheistic propaganda. This religious revival could have weakened the emerging idea of the Belarus nation, since the Belarusians had been traditionally divided between the Orthodox and the Catholic branches of Christianity. Given the history of Eastern Europe, and its realities that made the presence of secular leaders and intelligentsia so thin on the ground that the poor were often led by religious leaders, the dangers to Belarusian nationalism could become very real. However, Bekus's study shows that the ideology of Belarusian nationalism was possibly embedded in a very special kind of modernity. Instead of the dichotomy between religion and nationalism perceived in the Western discourse of modernity, in which they are positioned as opposing binaries, religion became a factor facilitating a shared sense of national belonging in Belarus. This acceptance of religion was heightened by the fact that it was one ideological system of historical continuity which had not been tainted by association with socialism. It is in this context, as a new nation endeavours to formulate its policies, that the author highlights how religious framings, though contested, of national identity could serve as a basis for sub-national 'groupness'. Nationalism – an ideology akin to an integrative belief system, similar to religion – can potentially be strengthened by the discourses of religion, making this composite discourse a valuable tool in constructing Belarusian national identity and papering over differences.

The second essay, by Keri Miller, explores the use of the term *ad-dīmoqrāṭīya* ('democracy') in Arabic discourse to reveal language as a politicized, non-neutral medium. Words that have come to straddle the world, such as 'democracy', pass with ease from one language to another. Though the users mistakenly believe that the word has of the same meaning across languages, in reality the word may inhabit an ambiguous zone within which it moves freely and with shifting associations. Miller examines the use of the term in a collection of articles, YouTube videos, and internet posts over the last ten years, to indicate how the contested association between Islam and the concept of 'modernity' may be the reason why the term 'democracy' has become a wild card in the discourse on Islamic movements. Used in connotations as diverse as (a) a potential solution to violence and a key to co-existence, (b) a threat to divine order, and (c) an empty promise employed as a cover for political and cultural hegemony, the word may be viewed as exemplifying terms which carry semiotic ambiguity reflecting cross-cultural tension. Miller highlights the contradictory usage of the term, indicating the possibility of manipulation of a word that is located within other extra-linguistic oppositions laid out in the discourse. Choosing articles at the forefront of Google searches and one YouTube video of a popular Al-Jazeera programme, she investigates the fraught relationship that the word's usage reveals in persons of varying ideological inclinations, pro-democratic and otherwise. This illustration may be extended to other Arabic terms, such as *at-taḥrīr* ('liberation') and *al-iṣlāḥ* ('reform'), terms that usually exist in association with democracy in a culturally loaded context.

The final essay, by Epsita Halder, studies the process by which the emerging Muslim middle class in Bengal sacralized modernity, making it possible for Muslims to be both Muslim and modern. Through a study of the discursive projects of individual Muslim literati in the late nineteenth and early twentieth centuries, Halder shows how writers connected the scriptural and the modern generic literary forms of 'biography' and 'history', the elite and popular sensibilities, and the sacred and the trivial. She argues that this flexibility of perception also succeeded in giving to Bengal an Islam that today is multi-linear and multiple. In this endeavour, Bengali Muslims were also writing a counter-narrative to that written by the north Indian Urdu-speaking intelligentsia and to that written by the Hindu Bengalis, who were socially more powerful.

This volume does not propose a single, seamless thesis on how languages and religions impact each other, and the purpose of using case studies is to base the themes in the specificities of history and in ethnographically researched locations. Drawing on ethnographic research, the chapters seek to analyse the dynamics of rapid social change resulting from linguistic and religious changes. Despite the width of geography addressed in the research, it has not been possible to include North America, North Asia, Central Asia, and much of Africa in this book. In fact, the book may even lose some salience by not being region- or era-specific. However, the objective being to explore the languages of religion in their persistent and widespread dynamics, a focused thematic work on one region or one era would not have illustrated the general nature of the subject. The volume respects and attempts to represent the heterogeneous nature of the material it examines – and it was not our intention to attempt to define a universalizing thesis in this area. The attempt, rather, has been to show that the margins can instigate changes and be empowering, though this does not preclude the centre as a place of power. In keeping with the current research on ethnicity and identities, the essays query rather than take for granted the religious and linguistic paradigms of the groups studied.

References

Anderson, Benedict. *Imagined Communities: Reflections on the Origin and Spread of Nationalism*. London and New York: Verso, 1991.

Green, Nile, and Mary Searle-Chatterjee. *Religion, Language and Power*. London and New York: Routledge, 2008.

Hobsbawm, Eric. "Introduction: Inventing Traditions". In *The Invention of Tradition*, edited by Eric Hobsbawm and Terence Ranger. Cambridge: Cambridge University Press, 1983.

Hoffman, Bruce. *Inside Terrorism*. New York: Columbia University Press. 2006.

Hoyt, Sarah F. "The Etymology of Religion", *Journal of the American Oriental Society* 32, No. 2 (1912): 126–9.

Kar, Bodhisattva. "Imagining post-Indian histories". Accessed 8 August 2015. www.india-seminar.com/2003/524/524%20bodhisattva%20kar.htm

Marx, Karl. "Introduction". *A Contribution to the Critique of Hegel's Philosophy of Right. Deutsch-Französische Jahrbücher*, 7 & 10 February 1844, Paris. www.marxists.org/archive/marx/works/1843/critique-hpr/intro.htm

Muller, Max. *Lectures on the Origin and Growth of Religion as Illustrated by the Religions of India: Delivered in the Chapter House, Westminster Abbey, in April, May, and June, 1878*, Volume 959. New York: Scribner's Sons, 1879.

Part I

INHERITED LANGUAGE TRADITIONS

1

CULTURES OF SOUND

Lineages and languages of sutra recitation in Goshirakawa's Japan

Charlotte Eubanks[1]

The seventy-seventh in a line of emperors tracing their origins to the gods, Goshirakawa (1127–92) spent a relatively short time on the throne, reigning for a scant three years (1155–8), but he continued to wield considerable influence as a "cloistered" or "retired" emperor until his death. His political legacies are myriad, and indeed the eruption of the samurai as powerful players in the central government can be traced to the years surrounding Goshirakawa's ascent to, and attempt to maintain, power. Eventually, Goshirakawa failed in his bid to retain the aristocracy as the political leaders of the country. In the area of cultural capital and soft power, however, Goshirakawa remained unrivalled, and his devotion to the vocal arts, the mastery of which he deployed repeatedly, is well-attested. As a young prince with "no immediate regal prospects" – he ascended to the throne largely because his predecessor's chosen heir died young – he "exploited his princely prerogatives to the fullest" (Kim 1994: 3), mastering several instruments (particularly the flute) and a wide variety of vocal musical genres, amongst them *saibara* (a sort of folk song), *rōei* (the recitation of Chinese-language poetry), *imayō* (a type of popular ballad), and the art form this chapter takes as its focus: *shōmyō* (the musical recitation of sutra segments).

Contemporary scholarship has focused mostly on Goshirakawa's interest in *imayō*, perhaps for two main reasons. First, he led the effort to produce written records of what had, until then, been a completely oral art form; he apprenticed himself for ten years, secured his place in a performance lineage, and composed two works: the *Ryōjin hishō* (Treasured selections of superb songs) and the *Kudenshū* (Collections of oral transmissions), thus preserving in writing a fragile oral tradition. Second, Goshirakawa's work opened up new vistas onto an art form mostly associated with lower-ranking or unranked women. While *saibara* and *rōei* had long been part of aristocratic music and thus were accorded a certain amount of gravitas, *imayō* was a less elite art form, many female practitioners of which were putatively involved in sexual trade (as seems to have also been the case with many male performers).

Goshirakawa's devotion to *shōmyō* was no less intense, and, in fact, his engagement with the musical recitation of sutras shares certain characteristics with his attentiveness to *imayō*, insofar as he extended his purview beyond what had properly been aristocratic art forms, attempting first to master the art and then to standardize it and establish himself firmly in a lineage of transmission. Shiba Kanoyo argued that Goshirakawa's attention to *shōmyō* was pivotal in transitioning it from an activity that monks engaged in as a spiritual, temple-based praxis to an artistic pursuit that aristocrats engaged in as entertainment at banquets and other social occasions (Shiba 1998). The transformation of *shōmyō*, of its language, and of its social uses is thus part and parcel of a larger narrative concerned with the medieval Japanese enculturation of Buddhism and the deployment of the vocal arts as a form of political soft power.

In the early centuries CE, Central Asian monks rendered the Buddhist scriptures into a host of languages, including Chinese. These monks chose to translate some terms while deciding to transliterate others. For instance, the Sanskrit word *śūnya* ('emptiness', as related to the interdependence of all phenomena) was often translated into the Chinese *ku* ('empty', 'open', or, as a noun, 'sky'), whereas *pāramitā* (the 'perfections' of an enlightened being) generally was transliterated as *boluomiduo* (four symbols whose basic meanings are, respectively, 'wave', 'silk gauze', 'honey', and 'many'), which entered the Sinitic lexicon as a loan word. It is one of the great curiosities of East Asian Buddhism that, even into the modern era, the sutras have remained, by and large, in the language of Classical Chinese. This historical peculiarity of the language of religion raises questions about how clerics and lay devotees conceived of and accessed the scriptures. As an exploration of these questions, this chapter will sketch the contours of recent Japanese scholarship on the musical recitation of sutras in and around the Japanese court and engage directly with a number of recently discovered musical treatises.

In many ways, this chapter builds on research I conducted for my first book, *Miracles of Book and Body: Buddhist Textual Culture in Medieval Japan* (Eubanks 2011). In that book, I primarily dealt with questions of reading, of the ways in which medieval Japanese people made sense of Buddhist religious texts which remained un-translated for centuries. I argued that the act of reading Buddhist materials in medieval Japan was perhaps more about performing a vocal text, about producing the sounds of sacred scripture, than it was about making grammatical sense of it. I based my thesis on the exploration of a vast genre of medieval literature (Jp: *setsuwa*, 'explanatory tales') for clues they give us as to what people meant when they said they had 'read' a Buddhist sutra.

In this chapter, I want to think about the same time period – medieval Japan (roughly the ninth through thirteenth centuries) – and about the same general set of issues – what it is that people meant when they said they had 'read' a Buddhist text. But instead of investigating *setsuwa* tales, I will consider a different textual archive: that of esoteric musical treatises and other documents pertaining to secret lineage transmissions. Accordingly, I will delve much more deeply into questions of acoustics, issues of sound and of music, and the ways in which making those sounds, creating that music of reading, became a recognized art form associated not only with major monastic centres but also with the imperial court.

I argue that medieval acoustic and performance-based practices in Japan – and their attendant literature of codification (manuals, secret transmission records, and so forth) – opened an interlingual space in which the tones of Chinese were dropped, without syntax being altered to fit Japanese semantic norms, such that semantically meaningful sound in Chinese was derailed without semantically meaningful sound in Japanese necessarily becoming entrained. This, however, was not a complete emptying out of meaning. Rather, by dwelling on, swelling out, breaking through, and melismatically extending the sound of each phoneme, the performers turned the incommensurability of language with meaning into the object of an extended, vocalized meditation, which itself became appropriated, as cultural capital, by the imperial court.

The languages of Buddhism in Japan

Scholars of Buddhism have long argued that "Great Vehicle", or Mahāyāna (the type of Buddhism generally dominant in East Asia), understands the sutra as a site of living presence (Abe 1999; Campany 1991; Gjertson 1981; Gummer 2005; Nakao 2001; Ruppert 2000). As the *Lotus Sutra* puts it:

> Wherever [this sutra] may be preached, or read, or recited, or written, or whatever place a roll of this scripture may occupy, in all those places one is to erect a stupa of the seven jewels, building it high and wide and with impressive decoration. There is no need even to lodge a *śarīra* [relic] in it. What is the reason? Within it there is already a whole body of the Thus Come One.[2]

In other words, sutras provide the seed that, properly nurtured, can transform a normal human into an enlightened being. On this understanding, sutras are a generative site, presumably capable of producing an unlimited number of buddhas. Any place that a sutra has existed (even if only as a passing vocalization) is thus a relic in its own right and is capable of producing miracles.

Considerations such as these led Gregory Schopen to argue, famously, that Mahāyāna Buddhism is a "cult of the book:" that devotion to the material form of sutras (scrolls, for instance) fundamentally defines Mahāyāna Buddhism and marks it as distinct from earlier Buddhist traditions (Schopen 2005: 25).[3] Indeed, Mahāyāna Buddhist texts repeatedly urge believers to engage with the sutras, for instance by reading them aloud from a written score or reciting them from memory, and they promise any number of rewards to those who do so. Some of these rewards accrue to the individual devotee. The *Nirvana Sutra* promises boons such as increased longevity (T 12.374.382c25) and even suggests that belief in the sutra can result in enhanced strength and good looks (T 12.374.399b5–7). The *Lotus Sutra*, most loquacious in this regard, claims that it shall reward its devotees with breath that never stinks, a mouth and tongue never diseased, teeth ever straight and white, a full and round face, and a perfectly formed male member, amongst other things (T 9.262.47a8–19).

Other potential boons extend beyond the individual, promising protection of the state, which is one reason why the Japanese court (like governments across South and East Asia more generally) sponsored the building of convents and monasteries and funded the duplication and recitation of particular sutras like the *Golden Light Sutra*, the twelfth chapter of which goes into great detail about the evil fate awaiting bad kings and the wonderful rewards accruing to good ones. In Japan, as early as 741 CE, Emperor Shōmu decreed the construction of temples in each of his provinces, at the head of which was the Tōdaiji in Nara, with its enormous statue of the Buddha. Across East Asia, the copying and recitation of sutras was a crucially important devotional activity for Mahāyāna Buddhists, a tendency which is only strengthened by the fact that attentiveness to accurate reproduction of grammar and syntax had been recognized as one of the core Buddhist intellectual disciplines even before the development of the Mahāyāna on the Indian subcontinent.

Reading the sutras, however, was neither a simple nor a straightforward proposition in medieval Japan, where the linguistic situation was almost impenetrably complex. The first sutras entered Japan in the form of Classical Chinese-language translations, and so they have remained, despite an awareness of both the presence of Sanskrit originals and the capacities for translation into Japanese. This situation has impacted the textual culture of sutras in Japan in several ways. Perhaps most immediately, in a practical sense, it entails a considerable gap between sound and sense. George

Tanabe (2004: 137) describes the situation succinctly, noting that the chanting of Chinese-language sutras in Japan

> produces sounds that cannot be recognized as regular spoken language. The *Heart Sutra*, for example, is popular in east Asia as a Chinese text about emptiness, a fundamental Mahāyāna teaching, but when it is chanted in Japan, each Chinese character is given a Japanese pronunciation without any change in the Chinese grammatical word order of the text. The audible result is neither Japanese nor Chinese, but a ritual language unto itself.

To 'read' a sutra aloud, therefore, requires one to vocalize a phonologically Japan-icized transliteration of a Chinese translation studded with Sinitic transliterations of Sanskritic loan words: an intensively hybrid and multi-glossic situation which, despite its complexity, has remained highly stable.

At the same time, Japanese Buddhists expressed an uncomfortable awareness of the realities and shortcomings of translation: crucially, the awareness that translation alters the acoustic character of a text, thus perhaps compromising its ritual efficacy, in that one no longer has access to the "word of the Buddha" (Sk: *Buddhavacana*), that is, the same sounds that came out of the historical Buddha's mouth when he was preaching (Eubanks 2013).

The upshot of this complex situation is that, in medieval Japan, as elsewhere in the Buddhist sphere, strong spiritual and secular motives almost mandated the creation of a recognized class of skilled readers (Jp: *nōdoku*, 能読); that is, ritual specialists whose area of expertise lay in mastering, performing, and eventually transmitting to the next generation acoustically accurate, ritually effective vocalizations of sacred text. Thus we see, particularly in the first centuries of Buddhism's enculturation on the Japanese archipelago, an intensive focus on phonology, a deep attentiveness to issues of pronunciation, perhaps most obviously in the form of monks whose specialty was *shittan* (悉曇, Sk: *siddham*); that is, Sanskrit grammar and pronunciation (Kanazawa Kenritsu Kanazawa Bunkō 2006a).

Defining *shōmyō*: from grammar to music

I claimed, a moment ago, that attentiveness to accurate reproduction of grammar and syntax were recognized, very early on, as one of the core Buddhist intellectual disciplines. Where is that claim rooted? The Sanskrit commentary *Yogācārabhūmi-śāstra* ("Discourse on the Stages of Yogic

Practice", composed circa 350 CE by the monastic scholar Asanga) lists five categories of scholarship. The first of these is *śabda-vidyā* (Jp: *shōmyō*, 声明, literally 'the illumination of voice'), which here can be generally understood as linguistic mastery of grammar and syntax.[4] Like a vast number of Buddhist texts, the *Yogācārabhūmi-śāstra* was translated into Chinese – the 648 CE translation into 100 volumes is perhaps the most important one – and its orthodox scholarly categorizations went on to provide textual justification for the development of intensive Buddhist focus on issues of grammar, syntax, and phonology in medieval East Asia, all of which is loosely termed *shōmyō*.

In Japan, the earliest instance of *shōmyō* being performed is the 752 CE eye-opening ceremony at Tōdaiji, in which, it seems, a large monastic group performed a number of vocalizations of Buddhist liturgical chant. Over the next five centuries in Japan, there was a marked shift away from any attempt to recreate the sounds of Sanskrit – or even of Chinese – and instead a movement toward vocal styles that emphasized increasingly nativized tones, rhythms, and syntax. In other words, musical recitation provided a space in which to experiment with *adapting* the sutras into Japanese, without ever *translating* the text in a word-for-word sense. (Indeed, not until the late nineteenth century was there a concerted attempt to produce a word-for-word translation of the Buddhist canon in a Japanese vernacular.) While *shōmyō* texts continue exist in Japan in a variety of written languages – Sanskrit, Chinese, and Japanese – I will be arguing here that, in fact, they would be best understood to occupy an interlingual register not wholly commensurate with any single language.

Reading a *shōmyō* text

Shōmyō texts utilize a complex system of musical notation called *hakase* 博士, an umbrella term which refers to any type of gloss or vocalization cue written onto a text. *Hakase* are "usually described in musicological writings for Western readers as *neumes* (or *neums*)", indicating generally a system of graphic signs that are a mixture of "contour oriented [signs designed] to 'look the way they sound'" and "conventional abstract symbols . . . requiring extraneous support for their reading" (Markam and Wolpert 2010: 172–3). Strokes of changing angles represent changing pitches, and the starting-point of any stroke may be oriented to the north, south, east, or west of the written Sinograph according to a performer-visualized "neume-map" (博士図, *hakase-zu*) whose directions are associated with relative pitch names and, at times, with finger-holes on the Japanese flute (Markham and Wolpert 2010: 173).

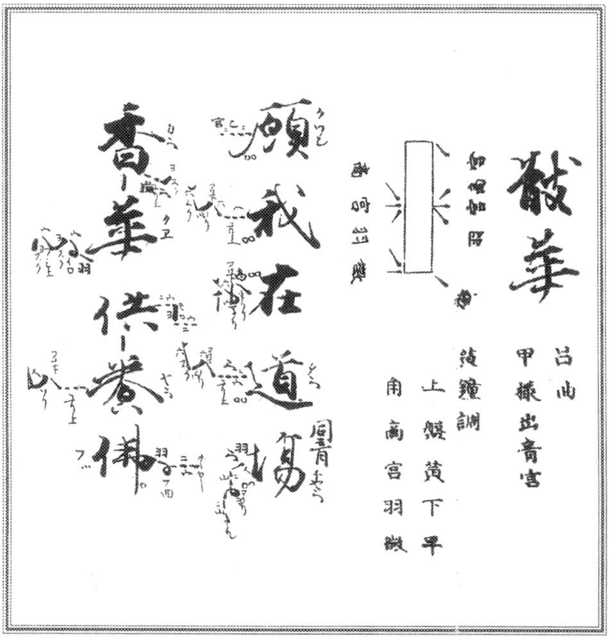

Figure 1.1 "Sange" (散華)

Note: Text for *shōmyō* recitation averaging about nine and one-half minutes, as currently performed at Raigōin (来迎院), a Tendai school temple to the north-west of Kyoto and locus of the Gyozan (魚山) lineage. To be recited during offering of flowers. Note in particular the '*hakase* map' (the long, narrow rectangle), as well as the extensive '*hakase*' pitch lines branching off to the sides of each of the ten words to be recited. The text reads: 'May it [always] be that I am in the place of practice, offering fragrant flowers to the Buddhas.'

Source: Reproduced from Ohashi 1999: 12–3.

These notations (Figure 1.1) originated largely in Japanese Buddhist contexts as monastics were struggling with how to vocalize Buddhist materials coming to them in the marvellously complex polyglossic situation I described earlier. Aside from using *hakase* to note which Chinese dialectal pronunciation to follow (Kan, 漢, or Go, 呉, regional pronunciations, for instance), monastics tried to come up with 'pointing' systems for marking the basic tones of spoken Chinese (commonly referred to now as rising, falling, dipping, and high tones). Over time, these tonal marks became melismatically extended, sometimes reaching incredible lengths such that the vocalization of a single syllable might take a full minute or even longer to perform (Markham 2008).

While the origin of *hakase* in Japanese Buddhist sutra recitation texts seems, then, to have been an attempt to provide tonal marks for speakers of the atonal language of Japanese, in time the notation system came to include a complex array of notational marks.[5] The most prominent of these are the "earthworm like" lines to the side of the written character, which are meant to visually represent both the melodic line and changes in pitch (Kanazawa Kenritsu Kanazawa Bunkō 2006b: 11). There was a great deal of variance across time and between schools of recitation, however, so the transmission of melody was functionally dependent upon master-to-student, face-to-face training. *Hakase* marks continue to be used in a wide variety of Japanese musical arts as the basic notation, even up to the contemporary era.

The consequences of mispronunciation

Having surveyed the general linguistic situation and some of the relevant cultural backdrop, let us turn now to a specific text, in order to get a sense of what is happening in these lineage treatises, most of which are marked as 'secret transmissions'. Shiba Kanoyo has created a typeset version of a document held in the Hōmyōin Temple Treasury. The document opens with a statement to the effect that all it contains is "deeply top top secret" (上之上之深秘也) and to be transmitted only within the lineage (Shiba 2004: 344). There are then four treatises transcribed one after the other: *Hokkekyō onzu* (法華経音図, A vocalization map for the Lotus sutra), *Darani-bon sōden* (陀羅尼品相伝, Transmission of the Dharani chapter [of the Lotus sutra]), *Suikunshō* (随訓抄, Notes on following Japanese syntax), and *Kangoshō* (閑語抄, Notes on things whispered). The text ends with a colophon dating the transcription to 1460, and noting that it is in turn based on a transcription from 1378, which is itself understood to be based on an oral tradition dating back to the twelfth century.

Let us examine a particular episode related in the *Kangoshō* (閑語抄), the last of the four treatises. The title of the document (*Kangoshō*) is already quite polyvalent and could be translated any number of ways – for instance, as 'notes on things best spoken about quietly' or as 'notes on correct language' – but the general sense of it is that what follows are secret transmissions, things which are only to be spoken about within a given lineage of initiates. Based on the title, we might surmise further that these notes will have something to say about the language of solitude or retreat and the language of religious devotion, as well as something to say about correct speech.

The *Kangoshō* opens by introducing the figure of Myōkaku (1056–1106), progenitor of a *shōmyō* lineage which claims Goshirakawa as one of its lineage-holders. Myōkaku was a scholar of Lotus pronunciation, achieving particular renown as a master of *siddham* (Sanskrit grammar) at Mount

Hiei, the major monastic centre just north of the imperial capital at Heian, before moving to Onsenji in Kaga prefecture later in his life. According to the text, Emperor Horikawa (1086–1107, r. 1079–1107) travelled to the district sometime in the Chōji era (1104–6). Finding himself unable to comprehend a single word of the local dialect, and aware of Myōkaku's reputation for being able to understand even the speech of birds and beasts, Horikawa summoned the monk, under whose direction he conceived an interest in learning the art of sutra recitation. Horikawa thereafter inaugurated an imperial lineage of *shōmyō*, which passed eventually to Emperors Shirakawa (1053–1129, r. 1072–86), Toba (1103–56, r. 1107–23), and finally Goshirakawa.

The *Kangoshō* also traces the Myōkaku/Goshirakawa lineage back to its point of origin, relaying a particularly colourful tale. One day, while walking down a path, Myōkaku is said to have come across a small snake being harassed by two young boys who were trying to kill it. In accord with his monastic vows to avoid harming living beings, Myōkaku chastised the boys and moved the snake to safety. A few days later, Myōkaku was walking down the same stretch of path when he met up with a handsome young man who asked the monk to follow him. The young man led Myōkaku down to the seashore, and into the ocean, escorting him to the palace of the Dragon King, who is recognized in Buddhist tradition as a protector of the sacred teachings. The young man then turned into a snake, summoned his father (the Dragon King himself) and related Myōkaku's act of kindness. Myōkaku and the Dragon King spent the night in conversation. Just before dawn, the Dragon King invited Myōkaku to choose one item from his textual archive to take back with him to Japan. The monk chose a volume on Sanskrit grammar, left the palace, and travelled back to Kaga province. Thus, the story suggests, there is an unbroken lineage from the Dragon King to Goshirakawa (Shiba 2004: 363).

The *Kangoshō* then provides a compilation of what can generally be thought of as cautionary tales related to the practice of musical sutra recitation gone wrong. There are general warnings, to the effect that mispronunciations may result in "filth issuing forth from your mouth" (Shiba 2004: 365) and may nullify any apotropaic boons. According to one of the stories in this section, the reigning Emperor Goshirakawa

> was reciting the [Lotus] sutra. When he came to the section in the fourth scroll with the four characters 燃大炬火, he pronounced the next to last character in the in the rising tone (上声) rather than in the correct tone. When he did so, everything became quite foreboding and, not even seven days later, the Imperial Palace

> caught fire. And thus we know that pronouncing the 炬 character in the rising tone brings misfortune, and that the correct pronunciation is to use the level tone (平声).
>
> (Shiba 2004: 365)

The sutra passage referred to here comes from book eleven of the *Lotus Sutra*, which is the "Apparition of the jewelled palace" chapter. What happens in this chapter is that the historical Buddha has gathered an assembly around himself and is preaching. All of a sudden, a wondrous, jewelled stupa rises out of the ground, and its doors open, revealing a previous buddha named Many Jewels (Prabutaratna). We hear that Many Jewels had made a vow such that, at any time an enlightened being preached the true dharma of the Lotus, his stupa would appear for all to see, as a miraculous sign of the veracity and authenticity of the teaching.

Why, though, should this little slip of the tongue set the imperial palace aflame? As it happens, the specific lines that Emperor Goshirakawa was chanting involve a metaphor. As part of the miracle being described in the sutra passage, a huge host of enlightened beings have appeared and gathered around the historical Buddha and, as enlightened beings are wont to do at key moments, each is emitting a light; each is literally glowing with the dharma. The sutra says, "Under each jeweled tree a different buddha sat on a lion throne, and from each emitted a wondrous light, blazing as a great bonfire in the dark of night" (其寶樹下，諸師子座 ，佛坐其上，光明嚴飾，如夜闇中，燃大炬火, T 9.262.0033c28–0034a01). What Emperor Goshirakawa did was slightly mispronounce the word "to blaze". This small error, we are given to believe, has incendiary repercussions.

Secret transmission and the 'way' of sutra recitation

The interlinguistic complexity of sutra texts in Japan, which is both clarified and complicated with the addition of *hakase* glosses, provides fertile ground for the cultivation of esoteric knowledge transmission. In a well-known and widely cited hypothesis, Konishi Jin'ichi argued that the notion that any given art form could comprise a 'path' (Jp: *michi*, 道; Konishi terms this an 'artistic vocation') to personal enlightenment or awakening (Jp: *jikaku*, 自覚) was a watershed cultural development of tenth- and eleventh-century Japan (Konishi 1991: 146–51). He lists five hallmarks of the transformation of an 'art' (Jp: *gei*, 芸) into a 'path':

1 specialization (dedication and focus on the part of the devotee)
2 transmission (establishment of a lineage of succession)

3 standardization (compilation of rules about how to perform the art, a 'conformist ethic')
4 universality (the 'cosmic view that an all-encompassing, universal truth dwells in even the tiniest, humblest of entities')
5 authority (development of patronage structures and hierarchies by which one's mastery might be recognized)

Konishi's argument achieves a theoretical accommodation between discourses of spiritual achievement and the logic of cultural capital, with exhortations to secrecy comprising one way of hoarding or preserving that capital.

Whereas Konishi's 'michi' concept focuses primarily on the classical and medieval periods, Maki Isaka Morinaga historicized the logic of esotericism across a broader spectrum, pointing out that "mechanisms of 'secret transmission', rather than 'secret information' per se" comprise the core logic of Japanese arts well into the modern era (Morinaga 2005, p. ix). Morinaga provided a five-stage schema, whereby the secret transmission (Jp: *hiden*, 秘伝) of a given art (*gei*, which he translates as "acquired technique") is first "systematized", generally in the context of esoteric Buddhism, is then "introduced to nonreligious communities" in a diffusion stage, is next "refined" and often codified in textual or set ritual format, and then is "commercialized" as status for sale before ultimately being challenged and rearticulated in the modern era (Morinaga 2005: 3). Konishi's concept of 'michi' correlates roughly to the period of transition between stage one (systemization) and stage two (diffusion) in Morinaga's model, whereas the surviving 'secret transmission' documents I am examining in this chapter generally belong to the third stage (codification).

To return to the story cited above, setting the imperial palace aflame seems a rather heavy weight to accord to a single-syllable error in pronunciation. Drawing attention to passages such as this, and connecting her archival resources to broader scholarly discussions concerning esoteric transmission in the Japanese arts, Shiba has hypothesized that prior to the mid-1100s, the practice of sutra reading in Japan was not very well codified. There were various "styles" (風) of reading (Shiba 2004: 29–30) – some involved reciting only the title of the sutra, some involved reading each word, some involved the use of percussive instruments to keep time, and so forth – and there was no standardization concerning how to actually pronounce the words (with or without tones, for instance). Various monastic centres had their own ways of doing this, and none of them seems to have been regarded consistently as any better than the other.

This rather laissez-faire situation began to change around the time of Emperor Goshirakawa. Goshirakawa's interest in sutra recitation as a musical form, according to Shiba, essentially tips the scales such that self-consciously codified schools of esoteric knowledge concerning the correct way to recite the sutras were founded. This trend started in the mid-1100s and continued to be elaborated over the next two centuries. A sense of "lineage consciousness" (*ryūha ishiki*, 流派意識; Shiba 2004: 116) developed, along with a certain rivalry and varying claims to authenticity, such that in relatively short order (within just 100 years or so) sutra recitation blossomed into a full-blown "path" (*michi*, 道) intertwining artistic mastery, spiritual pursuit, and political capital.[6] By the late 1200s, then, sutra recitation, on Shiba's account, had solidified as a recognized art form – the phrase "the way of sutra recitation" (*dokyōdō* 読経道) was first used in a document dated to the year 1284 (Shiba 2004: 1) – and instruction thereafter increasingly happened behind closed doors, within exclusive lineages which ended up producing a large number of 'secret transmission' documents, such as the one I have just explored.

Parsing the dharma: semantic vs acoustic units

In the story I examined above, Emperor Goshirakawa was performing a kind of sutra recitation called *kada shōmyō* (伽陀声明). This kind of recitation is based on the verse portions of Chinese-language translations of the sutras. One common feature of many sutras – most noticeably the *Lotus Sutra*, though it is a regular element in other major sutra texts as well – is to alternate sections of prose and poetry. In other words, sutra texts record an account of the Buddha giving a sermon in regular prosaic speech, and then there is a poetic restatement of what has just been said. These poetic recaps usually show up in highly regularized blocks, often of four-, five-, or seven-character phrases in the Chinese translations. In *kada shōmyō*, the practitioner recites four lines (*ku*, 句) of this poetry. The lines are given a standard melodic treatment when performed vocally.

Chinese-language translations of the sutras regularly feature poetic enjambment, however, meaning that a single semantic unit stretches across multiple lines or *ku*, so one cannot assume that any given line, as it appears as a phrase on the written page, will equate to the start of a semantically meaningful unit. This leads to lots of questions about where to draw the line, so to speak. To put it pointedly, how do you make sure that, when you are engaging in *kada shōmyō*, you are actually producing a vocal utterance that, however obliquely, refers back to a coherent grammatical unit? How do you know you are not reciting a sentence fragment?

Deciding where to divide *ku* thus became another matter of intense scrutiny in lineage documents and transmissions. Making this determination was quite difficult: there was no punctuation (such as commas or periods) to fall back on for clues, meaning that one had to trust that the founders of one's lineage had had enough facility with Classical Chinese to determine where sentences and phrases began.

There was another alternative – quite a radical one, in fact. Rather than reciting the sutra according to the text of a Classical Chinese translation, one could recite a poetic adaptation of it, a crystallization of the passage's essence as captured in Japanese-language poetic form (Gűlberg 1999; Kamens 1991; Miller 2013). This is exactly what we see in a late-fourteenth-century document from the *Shōmyōji* temple (称名寺) entitled *Hichūhi* (A secret amongst secrets, 秘中秘). The document notes that it represents the clarification or illumination of a teaching concerning the recitation of the Devadatta chapter of the *Lotus Sutra*, and then it opens with a *waka*, a thirty-one-syllable Japanese-language poem, that reads:

Transcription from fourteenth-century document:

> 法華経ヲワカエシコトハタキキコリナツミミスクミツカヘテソエシ
> (Kanagawa Kenritsu Kanazawa Bunkō 2006b: 8)

Transcription shifted to modern orthography:

> 法華経を我が得し事は薪こり菜つみ水くみ仕えてぞ得し

Transliteration into Roman characters:

> *Hokkekyō o waga eshi koto wa takigi kori na tsumi mizu kumi tsukaete zo eshi*

English-language translation:

> The reason I am now able [to preach] the *Lotus Sutra* is because [in a previous existence] I [served the sage Asita] by gathering his firewood, picking herbs for him, and carrying his water.

In the document, the poem is attributed to Emperor Kōmyō (1321–80, r. 1336–48), but the lyric was in fact composed by the Nara-era Buddhist monk Gyōki (668–749) and is preserved as poem number 1346 in the early-eleventh-century imperial poetry collection the *Shūishū*. For a *waka* it is rather bland and prosaic, but it fits into a relatively well-known subgenre of Japanese-language poetic composition called *shakkyōka* (釈教歌, literally "poem explaining the teachings of the Buddha" (Miller and Donnelly 2013). This particular *shakkyōka* attempts to capture the essence of the Devadatta chapter of the Lotus by zeroing in on the scene in which the Buddha reveals that in a past life he was a king who renounced his throne in order to seek spiritual insight. As a renunciant, he devoted himself to the seer Asita and served him for one thousand years; in return, Asita instructed the renunciant-king in the Lotus teachings. Many incarnations later, the renunciant-king has achieved enlightenment, been recognized as the historical Buddha, and preached the *Lotus Sutra*.

The poem, perhaps, is not particularly remarkable in and of itself, but it (and a host of poems like it) signals the dynamic linguistic malleability of sacred text in the pre-modern Japanese tradition. Aside from the title of the *Lotus Sutra*, the remainder of the poem is written in a native *katakana* syllabary, and to the left of each syllable extends *hakase*, providing notations of scale as well as indicating graphically the general melody. In other words, this *waka* is being recited by Buddhist adepts in the course of Buddhist ceremonies and is being transmitted through exclusive lineage instruction as a spiritually potent melody, with the understanding that its recitation may have apotropaic benefit. The *waka* is being treated as a sutra. In fact, this small poem is representative of an entire subgenre of sutra recitation known as *sandan* (讃嘆). *Sandan* is itself a particular variety of *kada shōmyō* in which the Chinese translations of the verse portions of the Sanskrit sutras have been completely and radically rearranged to fit Japanese syntax and poetic parameters.

Interlingual work is being done here, and it is the musical space that enables sacred text to move between languages. In musical recitation, a Japanese *waka* can *sound* like a sutra, can in fact become sacred text in its own right. There is a wonderful elasticity of translation going on here, an opening up of an interlingual space in which the tonal qualities of Chinese have been melismatically extended, hypertrophied into an intensive meditation on and through the vernacular breath.

Conclusion

As this discussion has shown, a great deal was at stake with sutra recitation in medieval Japan. Apotropaic boons are certainly part of the picture, but a close examination of secret musical treatises also shows clearly that in aristocratic circles and the often-overlapping monastic milieu of the eleventh through fourteenth centuries, the central concern attendant on sutra recitation may have had less to do with spiritual efficacy per se than with questions of cultural capital, whether in the form of placing oneself in a recognized lineage or of confidently producing a poetic vernacularization of sacred text. In fact, what we start to see by the 1100s is a move away from focusing on the pragmatic, protective value of sutra recitation (enacted by sponsoring and supporting others' readings of the sutras, for instance) toward considering sutra recitation as a personal art form, an aesthetic practice which one might master as a sign of individual status and accomplishment.

One promising extension of this research lies in further exploration of the ways in which both (a) the evolution of musical notation in Japan and (b) the evolution of terminology for describing acts of translation may be deeply intertwined with the history of how sutras came to be read. These broader implications await further research, and I hope to return to this fruitful field of enquiry in the near future. Examining the musical principles undergirding sutra recitation will enrich our terminologies for describing acts of translation. Ultimately, considering these texts may even enable us to move beyond the rather ubiquitous idea that the sutras were not translated into Japanese until the twentieth century. Instead, we might learn to speak more about the polyvocality of the situation. In translation studies terms, this would entail a basic recognition that we are dealing with a wide variety of translation practices, much more complex than a simple movement from source language (Chinese most often, Sanskrit less regularly) to target language (Japanese). Rather, the linguistic situation of sutras in medieval Japan can be seen to have comprised a massively complex blending of acoustic, syntactical, and melodic codes in which any notion of straightforward, linear movement of meaning from one language into another is problematized. *Shōmyō* thus provides one archive for 'sounding' the cultural depths of an inter- and multilingual Japan.

Notes

1 The author would like to thank Elizabeth Markham, Rembrandt Wolpert, and the University of Arkansas Departments of History and Asian Studies for the opportunity to think through many of the ideas conveyed in this article during a visit in the spring of 2013. Drs Markham and Wolpert have done much to improve my knowledge of musicology and of vocal performance in medieval Japan. Any errors or shortcomings, alas, remain my own.

2 As translated by Hurvitz, 1976: 178. The original text can be found at T 9.262.31b26–29. All quotes from sutras reference Takakusa et al. 1924–34, providing volume number, accession number, page, register, and lines. For example, T 9.262.31b26–29 refers to the twenty-sixth through twenty-ninth lines of the second (b) register of page 31 of Taishō text number 262 (the *Lotus Sutra*), which can be found in Volume 9 of the canon. Unless otherwise noted, all translations are my own.

3 As I have discussed elsewhere (Eubanks 2011), Schopen's thesis linking book culture to the origins of the Mahāyāna is an overstatement, and we should be wary of placing too much emphasis on it.

4 The other four areas of mastery pertain to technical craftsmanship, medicine, logic, and Buddhist doctrine. Markham notes that the term *shōmyō* "as first used in Japan was associated with the scholarly disciplines required of Brahmins in India. It referred specifically to the Sanskrit science of sound, to Sanskrit phonetics. Following Chinese terminology,

Buddhist ritual chant itself was called at first 'Sanskrit cantillation' (Jp. bonbai, Ch. fanbai). However, near the end of the Heian period (794–1185), the term *shōmyō* had widened its specific reference to include chant, eventually coming to replace the term bonbai. Nowadays, *shōmyō* is the general term for the ritual vocal music of Japanese Buddhism, and it covers texts written" in Sanskrit (known as *bonsan* in Japanese), in Chinese (known as *kansan*), and in Japanese (known as *wasan*) (Markham 2008: 11).

5 As musicologists have noted, the Japanese employment of *hakase* accompanied that culture's attempts, more broadly, to notate music from the Tang court in China (Picken et al. 1981–97). One description of these systems notes that "For the contrafactum, the transfer of the instrumental tune . . . into a neumatic notation, the written medium of its new Buddhist environment, carries the arsenal of support systems of the former instrumental occupant and lines them up with those that underprop the neumatic notation of the new resident. The tune-model is actually included as a base-melody in flute-tablature alongside the text-with-neumes. And a system of dotting that marks the metric grid for certain versions in an instrumental tablature is applied as glosses to the neumes, written above, below or to one or other side as support for melodic direction. The neumated Buddhist song itself presents as a vocal realization of the tune-model that appears to respond to its text-setting and vocal embellishment. . . . The neumatic system is essentially a graphic sol-fa, with directional single strokes and stroke-combinations with prescribed angles as pitch-indicators" (Markham 2004: 63).

6 A small group of Japanese scholars have, in recent years, begun to mine the two major archives of sutra recitation documents, the Kanazawa Bunkō near Tokyo and the Eizan Bunkō just north of Kyoto (Mabuchi 1985–92; Shiba 1998, 2001, 2004; Shimizu 1997, 1998, 2001). Though their work is far from complete, with new documents being discovered and transcribed each year, my survey of the scholarship produced thus far suggests no fewer than six major lineages The imperial lineage in the Heian capital (now known as Kyoto) was most active from the late 900s through the early 1200s. A second lineage, based around the monastic centre at Shoshazan, functioned from the mid-900s through the early 1700s and was at its peak from the 1200s through the late 1400s. Probably the oldest lineage, though it does not seem to have attained the same level of codification as the later ones, developed in the first stable imperial capital at Heijō (now known as Nara) by the 700s but did not achieve lineage consciousness, in Shiba's terms, until cross-fertilization with Shoshazan in the late 1400s. A fourth, more short-lived lineage seems to have been centred in the Musashi plains area, on the north-eastern frontier. The fifth lineage, purportedly founded by Myōkaku and centred around Mount Hiei, is marked by a great sense of rivalry with the imperial lineage, as is suggested by some of the materials examined in the body of this paper. A sixth centre of sutra recitation was founded by the monk Kūkai in the early 800s on Mount Kōya, though it seems to have engaged only minimally with, and indeed been only marginally interested in, the other lineages.

References

Abe, R. *The Weaving of Mantra: Kūkai and the Construction of Esoteric Buddhist Discourse*. New York: Columbia University Press, 1999.

Campany, R.F. "Notes on the Devotional Uses and Symbolic Functions of Sutra Texts as Depicted in Early Chinese Buddhist Miracles Tales and Hagiographies", *Journal of the International Association of Buddhist Studies* (1991): 28–72.

Eubanks, C. *Miracles of Book and Body: Buddhist Textual Culture and Medieval Japan*. Berkeley: University of California Press, 2011.

Eubanks, C. "Reading by Heart: Translated Buddhism and the Pictorial *Heart Sutras* of Early Modern Japan", *International Journal of the Sociology of Language* 220 (2013): 7–25.

Gjertson, D.E. "The Early Chinese Buddhist Miracle Tale: A Preliminary Survey", *Journal of the American Oriental Society* 101, no. 3 (1981): 287–301.

Gűlberg, N. *Buddhistische zeremoniale (kōshiki) und ihre bedeutung für die literatur des japanischen mittelalters*, Franz Steiner, Stuttgart, 1999.

Gummer, N. "Buddhist Books and Texts: Ritual Uses of Books". In *Encyclopedia of Religion*. 2nd edition, 1261–5. Detroit: Palgrave Macmillan, 2005.

Hurvitz, L., translator. *Scripture of the Lotus Blossom of the Fine Dharma (The Lotus Sūtra), Translated from the Chinese of Kumārajīva*. New York: Columbia University Press, 1976.

Kamens, E. *The Buddhist Poetry of the Great Kamo Priestess: Daisai'in Senchi and hosshin wakashū*. Ann Arbor, MI: University of Michigan Center for Asian Studies, 1991.

Kanazawa Kenritsu Kanazawa Bunkō. *Yomigaeru Kamakura no gakumon*. Yokohama: Kanazawa Bunkō, 2006a.

Kanazawa Kenritsu Kanazawa Bunkō. *Ji'in ni hibiku myō'on*. Yokohama: Kanazawa Bunkō, 2006b.

Kim, Y.-H. *Songs to Make the Dust Dance: The Ryōjin hishō of Twelfth Century Japan*. Berkeley: University of California Press, 1994.

Konishi, J. *A History of Japanese Literature, Volume Three: The High Middle Ages*. Translated by A. Gatten and M. Harbison. Princeton, NJ: Princeton University Press, 1991.

Mabuchi, K., ed. *Ei'in chūkai shittan gakusho senshū*. Tokyo: Benseisha, 1985–92.

Markham, E. "Contrafactum and a Buddhist Neumatic Notation from Medieval Japan", *Cantus Planus* (2004): 53–82.

Markham, E. "Chinese Hymns in Japanese Buddhist Liturgy: Structure and Ornament". In *Medieval Sacred Chant: From Japan to Portugal*, edited by M.P. Ferreira. Lisbon: Centro de Estudos de Sociologia et Estética Musical, 2008.

Markham, E., and Wolpert, R. "Developing a NeumeScribe for Sino-Japanese Buddhist Musical Notations", *Chung-Hwa Buddhist Journal* 23 (2010): 167–82.

Miller, S., and Donnelly, P. *The Wind from Vulture Peak: The Buddhification of Japanese Waka in the Heian Period*. Ithaca, NY: Cornell East Asia Program, 2013.

Morinaga, M.I. *Secrecy in Japanese Arts: 'Secret Transmission' as a Mode of Knowledge*. New York: Palgrave Macmillan, 2005.

Nakao, T. *Chūsei no kanjin hijiro to shari shinkō*. Tokyo: Yoshikawa Kōbun kan, 2001.

Ohashi, T. *Kyōto onfūkei: No. 2 shōmyō no katarai CD BOOK*. Kyoto: Shisuikai shuppan, 1999.

Picken, L.E.R. *Music from the Tang Court*, Volumes 1–6. London: Oxford and Cambridge University Presses, 1981–97.

Ruppert, B. *Jewel in the Ashes: Buddha Relics and Power in Early Medieval Japan*. Cambridge, MA: Harvard University Press, 2000.

Schopen, G. *Figments and Fragments of Mahāyāna Buddhism in India: More Collected Papers*. Honolulu: University of Hawai'i Press, 2005.

Shiba, K. "Dokyōdō kō", *Bungaku* 9, no. 1 (1998): 124–34.

Shiba, K. "Myōkaku to dokyōdō", *Bungaku* 2, no. 5 (2001): 149–67.

Shiba, K. *Dokyōdō no kenkyū*. Tokyo: Fuzama shobō, 2004.

Shimizu, M. "Nōdoku to nōsetsu: ongei 'dokyō' no ryōiki to tenkai", *Chūsei kayo kenkyūkai* 15 (1997): 22–36.

Shimizu, M. "Yomu kayō: shishū no ongei o musubu mono: *Dokyō kuden* o Shiten to Shite", *Nihon kayō kenkyū* 38 (1998): 54–63.

Shimizu, M. *Dokyō no sekai: nōdoku no tanjō*. Tokyo: Yoshikawa Kōbun kan, 2001.

Takakusu, J. et al. *Taishō shinshū daizōkyō*, 100 volumes. Tokyo: Taishō Issaikyō Kankōkai, 1924–34.

Tanabe, G., Jr. "Chanting and Liturgy". In *Encyclopedia of Buddhism*, edited by R.E. Buswell, Jr. New York: Palgrave Macmillan, 2004.

2

KAQCHIKEL SPIRITUALITY IN THE SHADOW OF CATHOLIC DOCTRINE

Andreas Koechert and Barbara Pfeiler

Latin America was broadly missioned and converted to Christianity since the Spanish and Portuguese conquest. The indigenous population of the colonies was exposed to Catholicism as state religion. This Latin-American Catholicism became the sole provider of social identities at regional and national levels.

The Maya of present-day Guatemala turned against the ideological-religious foreign domination of Catholicism and preserved many religious activities and behaviours of their autochthon culture. Despite the omnipresence of the Catholic Church, their spiritual traditions are maintained and lived by the traditionalists. Due to the lack of priests in rural parts of Guatemala, the responsibility of the Mission fell on an organization called *cofradía* (Catholic confraternity). According to Rojas Lima (1992: 271; author's translation), "the primary function of the *cofradías* can be described in a simplistic manner as a function of service and care of the saints, or as a permanent service activity for the community". The main duty of the *cofrades*, the confraternity members, has been to venerate their ancestors and divine beings, consult their priests, and visit holy places. In Guatemala, *cofradías* have become institutions where Maya spirituality is maintained and practised and have gained social autonomy similar to the local church communities. Thus, the confraternities developed a very special 'understanding' of how to treat sacredness, which is reflected in their prayers. Prayers are usually transmitted orally, so no written records have been kept (Gossen 1974; Brockmann 1991).

In San Juan Sacatepéquez,[1] a city of a small municipality with the same name and located in the Guatemalan highlands, several recent incidents endangered the maintenance of Kaqchiquel Maya spirituality (Koechert

2011), such as the earthquake of 1976 and the long "internal armed conflict" from 1960 to 1996 (CEH 1999). The religious-secular life amongst the Kaqchikel population of San Juan Sacatepéquez was determined by neoliberal economics; local, regional, and national policies; cultural particularism and universalism; and immanent ladino (mestizo) racism, but it was also influenced by the presence of the Catholic Church, protestant and evangelical sects, and the *costumbrismo*. For a long time, popular religiosity has competed with the official religion, which in broad strokes remains Catholicism.

In 1980, the confraternity members of San Juan Sacatepéquez expressed the need to preserve Kaqchikel traditional beliefs for future generations. Consequently, three *cofrades* selected, recorded, and wrote down a number of prayers and rites in Kaqchikel and then translated and interpreted them in Spanish with the technical assistance of Andreas Koechert, who did research at that time on cultural change in this town. The *cofrades* themselves added the annotations of the ceremonial context. This 'inner view' (Fishman 1997) facilitates the analysis of the prayers and helps to maintain the interpretation as faithfully as possible without leaving aside the limits of an overall intercultural communication. The *cofrades* required this work to be completed until after the end of the internal armed conflict. Furthermore, they requested the government to guarantee the free exercise of Maya spirituality. As it turned out, the earliest chance came with the peace treaty of 1996, and the knowledge of the informants has been edited gradually and carefully ever since.

This study will show how Kaqchikel spirituality has become resistant to the penetrating Catholic faith and its verbal doctrine. Kaqchikel spirituality manifests itself in ceremonial dances of pre-Hispanic origin, religious feasts, the veneration of Maya deities represented by the Christian saints, religious offerings and, above all, the Kaqchikel ritual language. The Kaqchikel ritual language can be understood as an expression of Christian and Maya faith intertwined with religious behaviour. Based on anthropological and linguistic data of Kaqchikel prayers, we focus on those specific words that show a mutual interaction between Catholicism and Kaqchikel Maya spirituality, in form as well as in content. In particular, we are interested in the intercultural semantics and the 'common' understanding achieved by the confraternity's sacredness. We consider that the highly structured texts, like the Kaqchikel prayers, are pertinent to show how religion affects language and language affects religion (Omoniyi and Fishman 2006) in the interaction of imported Roman Catholicism and native Maya spirituality. The importance of the role that language plays in the religious tradition of the Kaqchikel confraternity has been brought up in the understanding of the meanings that is defined by the proper cultural use given by three religious experts, namely the *prisipal* ('principal'), who is a high-ranking member of one of the five confraternities (García et al. 1998).

Religion and spirituality

Religion and spirituality are those activities and modes of human behaviour that are not related to the everyday life, the perishable, and the mundane. Today, the terms 'religion' and 'spirituality' are found in everyday discussions and in academic discourses. Generally, no definitions for these concepts are specified, but the meaning of religiosity and spirituality affect different people's purposes. According to Kippenberg (1999: 1174), religion can be categorized as follows: (1) intellectually, as a belief in supernatural beings; (2) functionally, as the power of social integration; (3) aesthetically, as the sentiment of a complete and pure dependency; and (4) convenient for life and in relation with special values.

Many scholars influenced the discussion of religion as a theoretical term. Amongst them, it is worthwhile to mention Edward Burnett Tylor, a contemporary to Charles Darwin and a representative of anthropological evolutionism; Robert Ranulph Marett, who marked the transition from evolutionism to functionalism through his scientific work on religion; Émile Durkheim, whose work is foundational to the definition of 'religion' from the functionalist point of view; and sociologist Max Weber, who differentiated the concept of religion from that of magic. Niklas Luhmann, with his systems theory, gave religion a functional orientation, while Clifford Geertz defined it as a symbolic system and a builder of cultures. Jacques Waardenburg determined religion scientifically as an occidental term and as inappropriate to describe religions from non-occidental cultures. Finally, some religious scholars suggest to replace the term 'religion' with terms such as 'faith', 'worldview', 'social formation', and 'culture'. Recently, there has been a scientific attempt to replace 'religion' with the term 'reflection', a way of reaching a scientific conceptualization through the conceptualization process itself. However, it is no longer the definition of the term, but the presence of religion as a 'voice', as a category, or as a concept, that is central to the discussion and academic interest (Alles 2005). The new stance determines religion as an 'open concept' that covers a range of functional and content determinations (Hock, cited in Heil 2010: 22).

Concerning the definition of 'spirituality', Sheldrake claims, "despite the fuzziness, it is possible to suggest that the word 'spirituality' refers to the deepest values and meanings by which people seek to live" (Sheldrake 2007: 2–3). A similar definition is found in Olupona: the "spiritual core is the deepest center of the person. It is here that the person is open to the transcendent dimension; it is here that the person experiences ultimate reality" (Olupona 2000: xii). MacDonald, on the other hand, highlights that "spirituality is the concern of human beings for their appropriate relationships with the cosmos. How the cosmic whole is conceived and what is considered appropriate in interacting with it differs depending on worldviews of individuals and communities" (MacDonald 2005: 8718).

In the context of Maya intellectuals, it is also difficult to find a clear meaning of 'spirituality'. Chile Pixtún (cited in Lima Soto 1995), Cojtí Cuxil (1997), Cabarrús (1998), Morales Sic (2004), and the members of Maya organizations such as the NGO Oxlajuj Ajpop (National Conference of Ministers of Mayan spirituality) or the Association of Mayan Priests of Guatemala (ASMG) have attempted to reach a 'common core' of understanding 'Maya spirituality'. These efforts are described in Morales Sic (2004), López Molina (2007), Ponciano (2007), and the Association of Maya Priests of Guatemala (ASMG). One particularity is the existence of a sacred universe including geographic locations, spirits of ancestors, hearts of heaven and earth, and others. For example, during rituals the believer directs himself or herself to the deities to ask for protection, support, and health, amongst other things. A 'set of conduct' or 'ethical register' (Ponciano 2007: 289) determines the coexistence of persons and provides references for the interpretation of reality. Additionally, there are experts in the sacred, who are, amongst other things, spiritual guides who mediate between believers and the deities. "All these elements form an ideological framework that feeds a sense of community and belonging" (Ponciano 2007: 289), which, according to the 'ideological' expression, would be defined as 'worldview'.

Amongst the Maya people of Guatemala, opinions vary about the definition of religious practices and spirituality. The term 'Maya spirituality' is preferred by persons who reject the term 'religion' because of its proximity to Christianity, particularly Catholicism. Moreover, they favour this term because 'spirituality' was coined by the Maya and has specific traits that distinguish it from Western religious thought.

The following example shows this difference: In the Kaqchikel community of San Juan Sacatepéquez, any alterations of spirituality based on the harmonic unity of the cosmos, nature, and human beings could be sanctioned by the deities and souls. The believer is afraid that spiritual harmony could be offended by his intervention, but his uncertainty is reduced and dissolved by the explicit link with the confraternity rituals. Straying from the spiritual rules is usually corrected in a normative way. That is, for instance, that public criticism of a person is much more relevant and constitutive for the people than the threats of sanctions and other coercive measures. Furthermore, the *cofrades*, particularly the principals, are dedicated to resolving problems of daily life, such as disputes between Kaqchikel people and even political problems.

The *cofradías*, or rather costumbrism, draws its strength from the ubiquitous preservation of the spiritual harmony of the cosmos, nature, and human beings. The spirituality underlying them outdoes the omnipresent Catholicism. On this basis, they have exerted considerable social influence within their community since their establishment during the colonial period.

Jean Molesky-Poz offers a deep and detailed knowledge of the spiritual and cosmological world of the Maya K'iche. At the same time, she points to the need for research on the sacred world in connection with reality when she says:

> Having defined spirituality as a practical form for living faith, ritual practice, and an encounter with Mystery, we must investigate who is addressed in ritual, that is, to investigate the models and metaphors of the sacred. From there we come to some assumptions about reality and the Creator-world relation. Some Ajq'ijab' pray to Catholic saints along with Maya epithets; other traditionalists invoke only names derived from a Maya construct. In prayer, Ajq'ijab' address the Mystery in various forms of couplets, each with its own valence.
>
> (Molesky-Poz 2006: 45)

This spiritual concept is characterized by the closeness to life, mutual exchange, and communality. The spirituality determines the life of the *cofrades* and *costumbristas.* Their "service of the Gods" is ritualized by their spiritual activity, but also by their daily work at home and in the field. Even if the term 'spirituality' is not comparable with the institutionalized term 'religion', there are intentions to establish it.

Kaqchikel prayers as sacred texts

The spirituality of the traditionalists and their organizations, whose representatives are the confraternities, play an important role inside the local religion.

Based on Flood (2006), we understand spiritual communication as a human practice limited by the perception of the sacred texts:

> Using the term 'sacred' does not, thereby, imply a reification of an area of human thought and activity such as to make it beyond inquiry or analysis, but it is to recognize that human communities mark off certain temporal, spatial, interior, and relational areas of life as having semantic density in contrast to mundane, transactional activities. Such areas are generally not idiosyncratic but determined by tradition and community.
>
> (Flood 2006: 52)

A sacred text may be formulated in writing or orally. In this manner, it

> pervades ritual, pervades the tradition's understanding and practice of economics, and pervades and shapes relationships between members of a society. Such a text-focused tradition creates a pattern of living and framework of values within which people live their lives in particular historical communities.
>
> (Flood 2006: 52)

In San Juan Sacatepéquez, the *cofrades*, who represent the spiritual elite of the local Kaqchikel *costumbristas*, focus their religious practices on the invocation of supernatural beings.

Confraternity prayers include the most ritualized forms of sacred expressions in the Kaqchikel community. They are regarded as the voices of sacred texts formulated by the same *cofrades*, which "are still resonant in those communities of transmission in the major traditions, and these voices make demands upon their readers, demands of repetition and liturgical performance" (Flood 2006: 53).

Confraternity prayers are an important element that structures spirituality. Their contents comprise the key terms of specific spiritual meanings and reflect the intellectual property of the *cofrades* and other Maya spiritual experts, such as diviners or priests. The *cofrades* use the prayers to communicate with transcendental beings and mediators (e.g. deities, angels, and saints). Catholic deities and saints, as well as indigenous deities, can be invoked in the same prayer, and the attributes of Christian saints can be assigned a differentiated interpretation in a *cofrade*'s contextual comment. Analogies are also drawn between Catholic doctrine and indigenous beliefs (see Koechert and Pfeiler 2013). This language allows

> a profound reflection of the comprehensive, symbolic system of indigenous spirituality and provides both the basis for expressing their coexistence within the indigenous community and family, and an instrument for protecting their thoughts and values.
>
> (Koechert and Pfeiler 2013: 133)

The starting point of this analysis is the finding that many key passages written in Kaqchikel include many religious concepts, terms, and words from the Spanish language. Standardized versions of Catholic segments and religious Spanish terms,[2] such as *wendision* (*bendición*), 'blessing'; *mister(io)*, 'mystery'; *gras(cia)*, 'grace'; *kastig* (*castigo*), 'divine retribution'; *permis(o)*, 'permission'; *sakrosant(o)*, 'sacrosanct'; and *milagr(o)*, 'miracle', are used according to their Christian meaning in Kaqchikel prayers. This integration of Spanish terms gives the appearance of following Christian doctrine and beliefs, as is shown by the following segment of the rain rite (rain-36), with the use of terms of the Trinity: the Father, the Son (Jesus Christ), and the Holy Spirit (1):

(1)

217	*táta señor Jesukrist*	father Lord Jesus
218	*ri kakläj táta*	the greatest father
219	*ri Dyos pagr*	the God Father
220	*ri Dyos hij*	the God Son
221	*Dyos espiritu sant*	God Holy Ghost

According to the *cofrades*, they perceive these words as being required by the dominant Catholic Church, and therefore in this context the prayers fulfil this duty.

In comparison with the K'iche' ritual speech (Rodríguez 2005), the integration of Catholic prayer segments in Kaqchiquel prayers occurs without intextualization. For example, the representations of the Christian saints, the Trinity, Maya divinities, or the *wendits anms* ('blessed souls of the ancestors') are no longer recognizable in their originality and Maya or ladino-Christian purity of character (see Watanabe 1990: 131–50); rather, they are authentic effects, attitudes, attributes, etc., that have been transposed and combined with each other in a given historical moment (Koechert 2007). The *wendits anms* (Sp: *ánimas benditas*) give shape, voice, thought, and history to their descendants, who are the actual living traditionalists of the Kaqchikel community. It is through *cofrades* that the blessed souls raise their 'voices' in form and content.

While the use of Spanish borrowings in *cofrades'* prayers represents the influence of Catholic thinking in Kaqchikel spirituality, the terms that are comprised by both the Catholic faith and Maya spirituality build the semantic background for the comprehension of the sacred. And even if some prayers appear to be Catholic due to their embroidering of the discursive forms (compare with Becker Richards 1988), the Christian content of the Spanish sacred terms is only partially integrated into the Kaqchikel prayers.

Let us take the use of the Kaqchikel term *mak*, which, translated into Spanish as *pecado original* ('original sin'), does not share the inherent meaning in the Catholic faith system. Instead, the word 'sin' is used in its original Kaqchikel meaning, which is 'a failing and a big spiritual mistake'. According to the *cofrades*, a person, or a group of persons, who have committed this 'fault' should actively correct it with effort and perseverance throughout a given period. In other words, a *mak* can be corrected only by taking a great punishment upon oneself in the present, the here-and-now in the world. A *mea culpa* as offered in the ladino culture would not be sufficient, just as waiting for a pardon would not suffice either.

This case shows how little the Catholic concepts of repentance and forgiveness have in common with Kaqchikel spirituality, and it also illustrates the lack of a proper term. The Spanish term used is *purtorio* ('purgatory'), or the word *ch'ajob'äl* ('washing', 'punishment'), or, even more interesting, the composition of the words *tojb'äl mak* ('payment', 'price', 'value', and 'big mistake'). All three terms denote the process of 'fulfilment' of punishment because of a fault and a big spiritual mistake (*mak*) in this world. If for any reason a man has an accident, for example, he may correct his mistake (*mak*) at that time. Thus purgatory is not the place (*purgatorium*) where the souls of the deceased are prepared for heaven in a process of purification, as is stated in Roman Catholic doctrine.

Other secular expressions apparently date back to the colonial era, the time of the founding of the *cofradías.* The following sequence (2) is taken from the rite of dance of moors (mor – 102) and refers to societal aspects of the past.

(2)

204	*ri ajaw Marks*	the Lord Mark
205	*ri ajaw Chepeton*	the Lord Chepeton
206	*ri Juan sik'ij*	the John Apazote[3]
207	*Manuel sik'ij*	Manuel Apazote
208	*ri Dyeg Martins*	the Didacus Martínez (everyman)
209	*Luks Martins*	Lucas Martínez (everyman)
210	*ri komon Dyeg*	the community of Didacus (all/everyman)
211	*ri komon Juan*	the community of John

The sequence of this prayer apparently has no Catholic meaning. The content refers to colonial local authorities and their town (*ri komon Juan*). *Ajaw* refers to a political title of the Maya elites in pre-Hispanic times. *Sik'ij* is the name for *criollo*, high-born Spanish colonists and their descendents. In colonial times, local and regional Spanish authorities received this denomination by the Maya population. The term *Manuel sik'ij* in the combination of 'Manuel' and the holy plant 'Apazote' refer to protection against the enemy. *Ri Juan sik'ij*, 'John Apazote', describes protection by the local patron saint, John, and the holy plant 'apazote', but no specific explication was given by the informants of the kind of protection. *Didacus Martínez* and *Lucas Martínez* mean 'everybody' of all Maya communities, whereas 'community of Didacus' are all Maya communities, and 'community of John' means the local Kaqchikel community.

Maya deities

The content of the religious words undergoes a transformation in the sense of a transignification embedded in the linguistic structure. The meaning of Catholic elements is changed and interpreted according to Maya spirituality. This concept follows the basic idea of the philosophical-theological concept of Jean de Baciocchi, Edward Schillebeeckx, Piet Schoonenberg, and other scholars of the modern Catholic theology. This concept is based on the idea of two kinds of presence, a local and a personal. In Catholic thought, Jesus is spiritually and personally present in the Holy Mass, but physically absent. This concept corresponds only to the Catholic elements of cofradial prayers, not to Kaqchikel spirituality, in which the divinity is both spiritually and physically present. Voss (2007) revisits transubstantiation from an older Catholic concept of Eucharistic Sacrament of the substance of Christ's Body and Blood:

> One of the fundamental aspects shared by these entities is the act of consecration. It is through sacralization that the transubstantiation operates by invoking the respective suprahuman entity, turning these material objects into material subjects. Thus inanimate matter becomes a living being that issues messages and acts on its own will.
>
> In the science of religions, we talk about fetishism and fetishes when the object that refers to a suprahuman entity becomes the same entity it represents. This can be seen clearly in the select cases that are introduced and analyzed, simply because the generic term used to refer to the inert object corresponds to the voice that denominates the suprahuman being; the divine entity personifies itself as substance and becomes the soul of the matter.
>
> (Voss 2007: 60)

The divine entity behaves humanly; this means, that despite being a material subject, it is alive but limited in its capabilities, which corresponds with the pre-Hispanic idea (concept) of Maya divinities. Following Voss (2007) and Kremer (2007, 2013) the transubstantiation is a unique component of Kaqchikel spirituality that is characterized by 'the service of and with the divinities'.

The Kaqchikel prayers of the *cofradías* contain a series of interesting analytical points, starting with the Catholic thought of sacred representations of divinity. According to Catholic theology, God alone is allowed to be worshipped; saints are not. Saints may be called only for the purpose of intercession with God. In popular religious practice, the boundary between veneration and worship of saints is fluid. These saints, as well as other divine beings, according to Voss (2007, author's translation), represent material subjects in the form of material objects in the spiritual thinking of the *cofrades* and *costumbristas*. They may be seen by them as "human beings". Wipf observes that saints "may lie, lose their composure, take revenge, have love affairs, and so on" (cited in Carlsen 2005: 5926). It is therefore not always appropriate to expect 'saintly' behaviour from them.

Kaqchikel divinities are seen as 'human beings' and behave as such. They are material subjects who manifest themselves in various representations (e.g. as Catholic saints or holy plants). As subjects, they are partners who have a mediating function as well as the very specific task to act 'in personam' to fulfil a request.

Catholic sacred images, which in the sense of transubstantiation or transignification experience a 'semantic transformation' from material objects to material subjects, possess an *anima de la materia* ('soul of matter') in the spirituality of the *cofrades*. They exist as both material objects and material subjects at the same time and without any transformation.

Locations such as churches, mountains, springs, and devotional objects and sacred times such as feasts and fast days also can be considered 'sacred', as long as it is proven that they serve to honour God.

Apparently Christian Spanish forms and figures of Kaqchikel prayers may have different sacred connotations than it seems at first glance. For example, material objects, such as the attributes of the saints, can be defined by their material subjects. Thus, for the *cofrade* or the *costumbrista*, the attribute of a saint is determinant for his veneration as a material subject.

Transcendence, which, according to Catholic doctrine, is associated with a saint, has little to no importance in this context. This transformation is characteristic of Kaqchikel spirituality and is exemplified in the prayer of Samarentin, taken from the rain rite (rain-03). Saint Martin's attribute is a monstrance with the sign of 'JHS' (Jesus Hominum Salvator) and a glorious sun inside. In fact, Saint Martin is the patron saint of farming, and his attribute is a wheel or a goose. In Spanish, however, this same saint is called San Bernardo, the patron of advertising, communications, and health problems. The principals explain that the name difference stems from the saint's attribute, which is the monstrance. In Kaqchikel spirituality, it corresponds with the iconographic representation of the sun's divinity. The principals describe these divergent names as follows: in their thinking, the naming convention is the monstrance of the saint, which concurs with San Bernardo de Siena, since the rays of sun coincide with the iconographic representation of the Kaqchikel sun god.

This shows that the figure, name, or status of venerability that a saint has in the Catholic cult does not correspond to Kaqchikel spirituality. For Kaqchikel believers, the iconographic attribute is determinant. What is spiritually relevant is the visual and tactile detectability of the attribute, not Catholic doctrine. Samarentin is, for Kaqchikel believers, 'touchable' as a material object-subject. As a deity, he lives a mundane existence amongst human beings. Clearly, the concrete and experienceable coexistence between the deity and the people determines the spiritual thinking of the *cofrades* and *costumbristas.*

Blessed souls

The *ánimas benditas* ('blessed souls', line 164 in example 3) are metaphorical representations of the deceased. They denote the person's state of being after death, in the *más allá* ('afterlife', 'afterworld'), in the *inframundo* ('underworld', 'afterlife') of the ancient Maya. According to the principals, the blessed souls of the afterlife and the *cofrades* of the present world are the mediators between people and the deities.

The *anm wendit* ('blessed soul') in the here and now is the Kaqchikel human who stays as such until his death. He gets back the status of a blessed soul, but only if he died of a natural cause. The blessed soul has its place also in the afterlife, with other blessed souls such as deities and angels who protect the human from the *ajliembre* ('damned soul', 'small beast') in the afterworld.

If the human dies of other than natural causes, such as an accident, murder, or suicide, in the afterlife he becomes an *ajliembre*, doomed to wander alone and live as a vagabond. The blessed soul returns as a human to the here and now by reincarnation. In the phase between birth and baptism, blessed souls forget all the knowledge they gained in their previous lives, both in this world and in the afterlife. All these forgotten memories characterize each individual's personality.

Furthermore, in the example there is another perception of the sacred term of 'soul'. It is *anms purtorio*, the 'soul of purgatory', which is the Catholic representation of the soul of the deceased in the purgatory and the anteroom to heaven. The following sequence (3) is taken from the *cofradía* rite (cof-96):

(3)

161	*ke re k'a*	thus
162	*xtinq'ajla' kichi'*	I'll upset their lips
163	*xtinq'ajla' kiwäch*	I'll upset their foreheads
164	*ri anms wendits*	the blessed souls
165	*anms purtorio*	souls of purgatory

The sacred term *ajliembre* does not appear in the prayers, but this character plays an interesting role in Kaqchikel spirituality, particularly in relation to the birth of a child. Unbaptized children are like little animals because, like animals, they do not have a place to stay. They never can relax, and they always have to go around looking for others to join them in their eternal life in the darkness. Therefore, the *ajliembres*, the deceased by spiritual mistake, try to approach unbaptized children with their destructive force of evil spirits to cause the children's death by illness, accident, or something similar. As for baptized children, on the other hand, the *ajliembres* have no right to harm them, because they became part of the good beings, the blessed souls, when they were baptized. Thus parents rush to have their children baptized in order to keep them from dying due to an *ajliembre*. Faithful mothers will not dare to take a child out of the house or leave him unattended until he is baptized.[4]

In comparison to the Catholic conception in which the blessed soul is understood as the incorporeal and immortal essence of a person, in Kaqchikel spirituality the blessed or the damned souls are the manifestations of humans in the afterworld. While the Christian conception is linear, with the soul that will be delivered from sin at its end through Jesus Christ (*krist Salvador*, 'Christ Saviour'; *Señor Munt Salwador*, 'Lord Saviour of the World'), the Kaqchikel conception refers to life as a cycle of birth, life, death, life in the afterworld, and rebirth. This life cycle has its equivalent in the annual cycle of sowing, growing, harvesting, and resting. The agrarian cycle is marked by feasts. The rural day cycle is largely dissolved and adapted to the requirements of the 'modern' work forms. Hence, it can be said that the prayers are a contrasting expression between the linear Christian and the circular Kaqchikel concept of time and space (Althoff 2005; Morales Sic 2004).

Maya deities are also invoked in the prayers. A sequence from the rain rite (rain-03) shows the invocation of the maize deity in five representations: Maria *toj*, Maria *tzame'y*, Maria *q'anel*, Maria *joko'*, and Maria *ch'umil* (5: 133–7).

(4)

122	*rix qate'*	you our mothers
123	*qa tata'*	our fathers
124	*táta señor*	Lord father
125	*chin kab'lajuj ik'*	for twelve months
126	*chin kab'lajuj ch'umil*	for twelve stars
127	*qub'in na wi*	we are always walking
128	*qu sutin na wi*	we always go around

129	*rix qate'*	you our mothers
130	*qa tata'*	our fathers
131	*táta señor*	Lord father
132	*Maria toj*	Mary Toj
133	*Maria tzame'y*	Mary Tzame'y
134	*Maria q'anel*	Mary Q'anel
135	*Maria jokow*	Mary Jokow
135	*Maria ch'umil*	Mary Ch'umil

For the *cofrades* and their believers, the *costumbristas*, the Kaqchikel cult of the maize deity has the same status and relevance as the opulent cult of the Virgin Mary amongst local Catholics. The connotations of *toj*, *tzame'y*, *q'anel*, *jokow*, and *ch'umil* show the spiritual importance of the maize deity.

This maize deity is "a complex structured potential divine being, who, in the form of divine humans, abstains from perceivable reality" (Koechert and Pfeiler 2013: 142). The deity is considered an anthropomorphic representation of maize. "[M]aize, [. . .] is the foundation of life and provider of strength and energy for human beings, influences almost all aspects of the man's life and teaches him the proper ways of connecting to the medium" (Arias 1975: 45; authors' translation). The connotations of the five sacred representations of the maize deity are considered allegorical descriptions, and they have the following meanings:

Toj indicates 'gift', 'offering', 'making a payment (for a fine)', or 'rain'. *Toj* is the 'corn cob', which allegorically means 'gift from the maize' with its bracts. In this way, *toj* represents the deity of the sun. It is also associated with the sacred fire and rain. Furthermore, the day *toj* in the Maya calendar is ideal for *cofrades* to make requests for others and receive advice from the divine authorities. In order for communication between the man and the deities to be successful, the former has to offer gifts to them. Within the agricultural cycle, *toj* is closely connected with the rain. *Tzame'y* makes reference to the maize sperm, i.e. its hair (silkcorn). *Q'anqel* is the 'germen' and 'seed', the 'greasy and fat grain of maize'. It symbolizes the spirit and allegorically means 'new life and creation', 'beginning' (germination), 'fertility', and 'abundance'. It also symbolizes the planet Venus, linked to origins and beginnings. It is associated with the seed, fertility, energy, fecundation, health, and food. The Q' sign in the Maya calendar represents, amongst others, the four major (grandparents), the compass (four cardinal directions), the four elements, and the four human races.

Jokow means 'reed cane' or 'mountain cane'. Baskets, used to transport and temporarily store the corncob, are woven from thin strips. Allegorically, *jokow* represents the guardian of the maize grains.

Ch'umil means 'star' and describes the 'star-maize' or fine grain maize from which (social and economic) development stems. It is in this germ that opens up at birth where the personal, familiar, and social development of each individual can be found. The term is used in the plural, *ch'umila*, in the context of birth or interaction.

The symbolism of maize in the Maya world is complex. It is worthwhile to mention that in Maya cosmogonic concepts the human being is made of maize: red maize corresponds to the blood, white maize to the bones, yellow maize to the skin, and black maize to the hair (Tixal Mul et al. 2001). As written in the Popol Vuh (Ximénez 1929–30), from maize the gods created human beings.

Another example is taken from the prayer to Saint Jerónimo (jer-02), which clearly shows the essential differences between the Catholic and the Kaqchikel expressions in a rite of prayers and the spiritual content of the Maya worldview (sequence 3: 20–6). It evokes the sun deity (3: 20–1), which, upon reaching the zenith, exerts its greatest influence and power. Its divine power is associated with the birth of the human being and the creation of the new. In this moment, the sun deity provides the *cofrade* Hieronymus (its 'companion' or interlocutor) with the necessary power to mediate the reading and the writing (3: 22–6). In this part, which is a central element of the prayer, the epiclesis, particularly the invocation of God, is nowhere to be found, as one would expect in the Catholic sense. Conversely, the relation is on one hand between the supplicant and a divine being (deity), and on the other between the supplicant and a *cofrade* who acts as a mediator between the human and the deity. The core message of the prayer therefore clearly belongs to Kaqchikel spirituality and not to Catholic doctrine. Such passages, in spite of historical efforts at ideological alienation by dominant Spanish and Catholic forces (which apparently did not reach their goal of dissolution of Maya spirituality), verify the originality and historicity of spiritual thought, because they date back to the pre-Hispanic time of the Maya culture.

New prayers, in the sense that they did not exist in the oral tradition, are created by the *cofrades* and are based on Catholic saints in accordance with the necessity of the supplicant. In the following prayer of the rite of learning (jer-02), its managerial function and the self-awareness of the *cofrade* are made explicit. The content is about the Kaqchikel people, who want to learn to read and write. In order for this wish to come true, some *costumbristas* approach a *cofrade* with the request to help them to fulfil their desire. The *cofrade* responds to this request with a spiritual ceremony and the following prayer.

(5)

(1)	*Señor San Jerónimo*	Lord Saint Hieronymus
(2)	*Señor San Jerónimo*	Lord Saint Hieronymus
(3)	*rat secretari mayor*	you are the main clerk
(4)	*rat secretari mayor*	you are the main clerk
(5)	*Señor Munt Salwador*	Lord Saviour of the World
(6)	*perdon chachi'*	forgiveness from your lip
(7)	*perdon chawäch*	forgiveness from your forehead
(8)	*táta señor*	Lord father
(9)	*xkenanuk'*	who shapes me
(10)	*xkenaweq*	who adorns me
(11)	*k'in r asamaj*	with my work
(12)	*k'in r apatan*	with my sacrosanct
(13)	*perdon chachi'*	forgiveness from your lip
(14)	*perdon chawäch*	forgiveness from your forehead
(15)	*komi or par nuk'ub'al*	today's time in the testament
(16)	*tzub'al*	in the past
(17)	*Señor San Jerónimo*	Lord Saint Hieronymus
(18)	*kas kinakuy nawi*	you who teach me
(19)	*k'in ra letr*	with your letters
(20)	*chin k'aj q'ij*	for sky of sun (noontime)
(21)	*k'aj säq*	white sky (middling clarity)
(22)	*tz'om rasamaj yin*	I will take your work
(23)	*achike mod xtiwatemaj*	that a way to learn
(24)	*xtib'än leer*	make to read
(25)	*xtib'än escribir*	make to write
(26)	*Señor San Jerónimo*	Lord Saint Hieronymus
(27)	*xa jun lisens*	as a license
(28)	*xa jun permis*	as a permit
(29)	*nk'utuj xe' rq'a'*	I ask you under your arm
(30)	*awaqän*	your foot
(31)	*komi q'ij*	today
(32)	*komi or*	time of today
(33)	*perdon chachi'*	forgiveness from your lip
(34)	*perdon chawäch*	forgiveness from your forehead
(35)	*pa rtzub'al*	in the past
(36)	*nuk'ub'al*	(in the) testament
(37)	*ri komi or*	the time of today
(38)	*Señor San Jerónimo*	Lord Saint Hieronymus
(39)	*komi q'ij*	today

(40) *komi or*	time of today
(41) *ja ri' k'utuj*	so this I ask you
(42) *xe' rq'a'*	under your arm
(43) *awaqän*	your foot

Regarding the creation of this prayer, principal Ciriaco García remarked that, being an elder illiterate, he once participated as a student in a literacy campaign. But the activity was so difficult for him that after only a few days, he decided he had to find another way of becoming literate. As may be expected from a *cofrade*, i.e. a wise man (*ajq'ij*), he turned to one of the many saints and asked for help. He evoked St Jeronimo, known amongst the Kaqchikel as the clerk who knew how to read and write due to his attributes, which are books and writing materials. The other attributes, a hat and the cardinal dress, a lion and, to a lesser extent, a cross and a skull that were considered St Jeronimo's primary attributes, were not important in this context. To his regret, as the principal himself commented, the ceremony and prayer did not have any effect and he remained illiterate. The *cofrade* set out to find a reason for his lack of success, to avoid hurting his public reputation. He spent a long time looking for a useful solution until he found it in a dream in which an old man dressed in white appeared. The old man said: "But he [the *cofrade*] did not learn how to read and write because, according to him, St. Jeronimo told him in a dream that learning to read and write was not worthwhile since his [the *cofrade*'s] job is to heal people or teach how to become a prisipal" (García in García et al. 1998; author's translation).

The *cofrade* presented the petition of the prayer before the altar to St Jeronimo, who had reached a step higher than the grace of God. He did not present the petition directly to God, as is usual in Roman Catholic doctrine. According to Kaqchikel spirituality, the *cofrade* is regarded as the most appropriate and capable mediator between the supplicant and God. St Jeronimo acts not as intercessor between the supplicant and God, but rather as intervener in the supplicant's life in a pleading way. In order for this to happen in reality, the *cofrade* offers his support through prayer (3; 22–5). As mentioned above, the *cofrade* regards the saint as a human being and treats him collegially and professionally. The fact that the *cofrade* did not have to learn to read or write is based on a dream in which St Jeronimo explained that a *cofrade* does not need to have such competence, for he is a wise and superior man due to his spiritual knowledge. Hence the *cofrade* did not have to be hesitant about the success of the implementation of the literacy programme. Yet he acted cautiously when he helped the petitioner who wanted his children to learn to read and write through the prayer and the ceremony. In this way the *cofrade* advised the petitioner to have his children attend school.

Linguistic style

Kaqchikel prayers encompass the typical Mesoamerican structure that show the high rhetorical style of a formal language comprehended by repetitions, parallelisms (Jakobson 1966; Hymes 1992; Sherzer 1990), diphrasisms, couplets, and dichotomic expressions (Bricker 1974, 2010; Gossen 1974; Hanks 2010; Hill and Hill 1986; Monod Becquelin and Becquey 2008; Montes de Oca 2004). The Spanish lexicon in the prayers is perfectly fitted into the Kaqchikel ritual speech structure, including the literary figures of parallelism and diphrasism, both present in Maya hieroglyphic scripts used since pre-Hispanic times (Lacadena 2010).

The prayers are characterized by an initial phrase of invocation and the repetition of this segment. Different types of repetition occur in the prayers: the same element is repeated at the beginning or the end of couplets. Repetitions can also be more separated, providing the effect of rhythm.

If repetition of the morphological element occurs in an initial position, the figure is called parallelism. Parallelism is a syntactic form of repetition that consists of distribution of a verbal message in two constituents composed by two elements, the first being an anaphoric relation and the second consisting of two terms in a semantic relation (Lacadena 2010: 58). In the following single parallelism (1) the pair B1-B2 maintains semantic relationship. The example belongs to the Hieronymus prayer (jer-02):

A-B_1
A-B_2

(1)

27 *chin***A***k'aj* **B**$_1$*q'ij* for sky of sun (for noontime)
28 **A***k'aj* **B**$_2$*säq* white sky (middling clarity)

An example of multiple parallelism (more than two constituents, in which the second elements have to be related semantically) is shown in (2), taken from the *cofradía* rite (cof-01).

A-B_{11} A-B_{12} A-B_{13}

(2)

51 **A***tiya' k'a ri* **B**$_{11}$*grasia* let them give grace
52 **A***tiya' k'a ri* **B**$_{12}$*wendision* let them give blessing
53 **A***tiya' (k'a) ri* **B**$_{13}$*sentid* let them give consciousness

In this example, *tiya* is the phonological repetition of the optative *ti-* with the verb *ya'* for 'give'. This is in an initial position in both constituents, while B_1, B_2, and B_3 are three formally distinct elements, which maintain a semantic relationship amongst themselves.

Compared to the syntactic repetition figure of parallelism, the stylistic literary resource diphrasism is a semantic substitution figure, i.e. a metaphor or metonym. Diphrasisms, embedded in a parallelism figure (3), occur in the rain ritual prayer (rain-03):

(3)

122	*Táta Señor*	Lord father
123	*kinuk'un na wi*	they have always sculptured
124	*kiweqon na wi*	they have always ornamented
125	*chin kab'lajuj ik'*	for the twelve moons
126	*chin kab'lajuj ch'umil*	for the twelve stars
127	*tib'e' q'ij*	the day passes
128	*tib'e' säq*	the clarity passes
129	*rix qate'*	you our mothers
130	*qa tata'*	our fathers

The pair *ik'/ch'umil* ('moons'/'stars') refers to the twelve months of the year, which in turn represents an association with the twelve apostles.

The high rhetorical style of formal language also includes dichotomic expressions, which are pairs of opposites that speak about location, spiritual values, and nature. Dichotomic expressions are often referred to as the characteristic feature of dualism in ancient and modern Mesoamerican cultures. Dualism has been considered a significant trait of indigenous Mesoamerican religions, although this remains open to debate. Modern Kaqchikel spirituality contains frequent dualistic representations, but it also involves deities and anthropomorphic divinities with multiple manifestations, exemplified in this study with the presentation of the maize deity (3: 123–7) and the relation between person and plant, i.e. Mary and corn.

Conclusion

The multiplicity of religions that are present in Guatemala greatly hinders the development of a structural outline for certain religions. Maya religiosities and identities are no exception to this, especially given the ephemeral distinction between the religious and the secular world. It is therefore difficult to develop a unique concept for the term 'Maya spirituality'. First, its existence and presence in every sector of the daily life is acknowledged.

During the internal armed conflict in Guatemala, many expressions related to Kaqchikel spirituality were in danger of being destroyed. Hence, recording their knowledge and commenting on it themselves was of utmost importance for the principals as spirituality caretakers. By keeping record of a number of Kaqchikel prayers, they managed to preserve and maintain Kaqchikel spirituality for themselves, the remaining members of the *cofradías*, the *costumbristas*, and finally, the Kaqchikel community.

The sacred texts formatted as prayers essentially evince the core of Kaqchikel spirituality. According the principals, prayers had to be conserved intact in structure and content to avoid altering the harmony of the cosmos, nature, human beings, and the deities. Human beings may be punished by the deities when they cause a disruption by failing with the ritual duties. However, the deities' punishment ability does not have to be necessarily exercised. Additionally, there is the fear of human interference with the existing spiritual harmony, an uncertainty that is reduced and dissolved by the explicit link with the cofradial rituals. Such deviation from the spiritual rules is usually corrected in a normative way. Public criticism of a person is much more substantial for the people than the threat of sanctions and other coercive measures (García et al. 1998).

Spirituality is made manifest through the worship of the deities, saints, and *animas* ('souls'), who are considered partners of the human beings by the *costumbristas* and the cofradial members. Deities and other divine beings are considered human beings who see, hear, smell, and feel like humans and who manifest themselves through prayers and ceremonial attributes such as *pom* ('copal'), candles, sacred liquor, food, coins, cigars, and more. It is by these means that the divine is manifested figuratively. It is an act of consecration in which the divinities are transformed into material subjects that are concrete representations, i.e. characters that are identical to the human being. The object receives its 'anima' of subject.

It is through the sacred texts, the cofradial prayers, that the *cofrades* and the *costumbristas* find and establish the quotidian link to their lives and nature, the cosmos, and the people. These essential elements are perceived, owned, and applied by the people in time and space that coincides with the rhythms of nature and the cosmos through some reciprocal recognition proceedings. Therefore, there is a dialogue amongst all the subjects of the world during the veneration within and through prayers: *cofrades*, *costumbristas*, *Kaqchikeles* on one side, and deities, saints, and souls on the other side. All of them are fundamental, integrated components of a whole, in which the divinity group serves humans and vice versa, each of them being fallible or 'human' and thus having rights and obligations. The influence of Catholic thinking is barely noticeable, but Kaqchikel spirituality is omnipresent and dominant.

The language of the sacred texts is revealed as a mode of expression that is native to Maya spirituality. Due to the Christian terms in Spanish which sometimes appear not only at the beginning of a sentence or couplet, but also within it, the prayers appear to be Catholic at first sight. Such terms may be omitted if the prayer is a stereotype, as in the case of the salutation. It can be said that, apparently, the real spiritual statement is wrapped in a Christian Catholic context in a kind of ideological and linguistic camouflage.

Kaqchikel spirituality is alive and challenging, an element of the religious and ideological identity that is native to the Kaqchikel *costumbristas.*

Notes

1 The region is in central Guatemala, in the rural highlands north-east of Guatemala City and covers 242 km^2. In 2006 it had approximately 193,000 inhabitants, of which about 85 per cent were Kaqchikel-speaking Mayas.

2 Most Spanish borrowings in Kaqchikel undergo phonological reduction and are adapted to the Kaqchikel stress pattern.

3 *Apazote* means 'holy plant', 'medicine', 'spice', or 'hot herb' and is the common name for *Chenopodium ambrosoides CS apazote.*

4 Another reason is health-related: the *comadrona* (midwife) forbids the mother to do any kind of hard work in or outside her home in order to avoid complications. In addition, it is uncommon for a Kaqchikel to see a mother leave the house.

References

Alles, Gregory D. "Religion [Further Considerations]". In *Encyclopedia of Religion*, Volume 11, edited by Lindsay Jones (Editor in chief), 7701–6. 2nd edition. Farmington Hills: Thomson Gale, 2005.

Althoff, Andrea. "Religion im Wandel: Einflüsse von Ethnizität auf die religiöse Ordnung am Beispiel Guatemalas". PhD thesis. Martin-Luther-Universität Halle-Wittenberg, Halle, 2005.

Arias, Jacinto. *El mundo numinoso de los Mayas: estructura y caminos contemporáneos.* México: SEP, 1975.

Bricker, Victoria R. "The Ethnographic Context of Some Traditional Mayan Speech Genres". In *Explorations in the Ethnography of Speaking*, edited by Richard Bauman and Joel Sherzer, 368–88. Cambridge: Cambridge University Press, 1974.

Bricker, Victoria R. "La función del paralelismo en la narrativa tzotzil". In *Figuras mayas de la diversidad*, compiled by Aurore Monod Bequelin, Alain Breton, and Mario Humberto Ruz, 87–100. México: Universidad Nacional Autónoma de México, 2010.

Brockmann, Andreas. *Santa Martha – Untersuchungen zur Ethnographie einer Tzotzilgemeinde in Mexiko.* Münster: Lit-Verlag, 1991.

Cabarrús, Carlos Rafael. *La cosmovisión Q'eqchi' en proceso de cambio*. Guatemala: Cholsamaj, 1998.

Carlsen, Robert. "Mesoamerican Religions: Contemporary Cultures". In *Encyclopedia of Religion*, Volume 9, edited by Lindsay Jones (Editor in chief), 5923–33. 2nd edition. Farmington Hills: Thomson Gale, 2005.

CEH see Comisión para el Esclarecimiento Histórico (CEH).

Cojtí Cuxil, Demetrio. *El movimiento maya en Guatemala*. Guatemala: Cholsamaj, 1997.

Cojtí Macario, Narciso, Martín Chacach Cutzal, and Marcos A. Calí Semeyá. *Diccionario Kaqchiquel*. Guatemala: P.L.F.M. – Cholsamaj, 2001.

Comisión para el Esclarecimiento Histórico (CEH). *Guatemala: Memoria del silencio*, 1999. Accessed 30 June 2013. http://shr.aaas.org/guatemala/ceh/gmds_pdf/

Fishman, Joshua. "Language and Ethnicity: The View from Within". In *The Handbook of Sociolinguistics*, edited by Florian Coulmas, 327–43. Oxford: Blackwell, 1997.

Flood, Gavin. "Reflections on Tradition and Inquiry in the Study of Religions", *Journal of the American Academy of Religion* 74, no. 1 (2006): 47–59.

García, Ciriaco, Francisco Sajquy, and José Maria Sequén. *Colección de oraciones cofradiales*. Recorded in San Juan Sacatepéquez, translated from Kaqchikel, and archived by Andreas Koechert. Manuscript Bremen, 1998.

Gossen, Gary. *Chamulas in the World of the Sun: Time and Space in a Maya Oral Tradition*. Cambridge, MA: Harvard University Press, 1974.

Hanks, William. "Tzol: Paralelismo cíclico en el Maya yucateco colonial". In *Figuras mayas de la diversidad*, compiled by Aurore Monod Bequelin, Alain Breton, and Mario Humberto Ruz, 479–506. México: Universidad Nacional Autónoma de México, 2010.

Heil, Joachim."Was Ist 'Religion' ? Eine Einführung in unser wissenschaftliches Reden über Religion", *Internationale Zeitschrift für Philosophie und Psychosomatik* 2010, no. 1 (2010): 1–29. Accessed 28 June 2013. www.izpp.de/fileadmin/user_upload/Ausgabe-1-2010/02_1-2010_TS_Heil.pdf

Hill, Jane, and Hill, Kenneth. *Speaking Mexicano: Dynamics of Syncretic Language in Central Mexico*. Tucson: University of Arizona Press, 1986.

Hymes, Dell. "Use All There Is to Use". In *On the Translation of Native American Literatures*, edited by Brian Swann, 83–124. Washington, DC: Smithsonian Institution Press, 1992.

Jakobson, Roman. "Grammatical Parallelism and Its Russian Facet", *Language* 42 (1966): 398–429.

Kippenberg, Hans Gerhard. "Philosophie XV: Religionsphilosophie/Religion". In *Enzyklope die Philosophie*, Volume 2, edited by Hans Jörg Sandkühler, 1174–83. Hamburg: Felix Meiner Verlag, 1999.

Koechert, Andreas. *Xe' Ruq'a', Xe' Raqän – Cargos cofradiales Kaqchiqueles*. Guatemala: Cholsamaj, 2007.

Koechert, Andreas. "La espiritualidad en oraciones Cakchiqueles". In *Ritos, culto y cosmovisión en Mesoamérica: pasado y presente*, edited by Alexander

Voss, Andres Koechert, and Yuri Balam Ramos, 95–117. Graz: ACPUB Academic Publishers – Universidad de Quintana Roo, 2010.

Koechert, Andreas. "Un pueblo Kaqchikel ante un conflicto armado ¿Un caso de autopoiesis?", *Ketzalcalli* 2011, no. 1 (2011): 3–24.

Koechert, Andreas, and Barbara Pfeiler. "Maintenance of Kaqchikel Ritual Speech in the Confraternities of San Juan Sacatepéquez, Guatemala", *International Journal of the Sociology of Language* 220 (2013): 127–49.

Kremer, Jürgen. "Religión: una definición para Mesoamérica", *Ketzalcalli* 2007, no. 2 (2007): 3–20.

Kremer, Jürgen."Religion heißt Dienst an den Göttern: eine kultur – und religionswissenschaftliche Untersuchung zu den Grundlagen des mesoamerikanischen Polytheismus. " PhD thesis, University of Hamburg. Hamburg, 2013.

Lacadena, Alfonso. "Naturaleza, tipología y usos del paralelismo en la literatura jeroglífica". In *Figuras Mayas de la diversidad*, compiled by Aurore Monod Bequelin, Alain Breton, and Mario Humberto Ruz, 55–85. México: Universidad Nacional Autónoma de México, 2010.

Lima Soto, Ricardo, E. *Aproximación a la cosmovisión maya*, Guatemala: Universidad Rafael Landívar, Instituto de Investigaciones Económicas y Sociales, 1995.

López Molina, Ana. "La Asociación de Sacerdotes Mayas de Guatemala". In *Mayanización y vida cotidiana. La ideología rnulticultural en la sociedad guatemalteca*, Volumen 2: Los estudios de caso, coordinated by Santiago Bastos, and Aura Cumes. Guatemala: FLACSO – CIRMA – Cholsamaj, 2007.

MacDonald, Mary. "Spirituality". In *Encyclopedia of Religion*, Volume 13, edited by Lindsay Jones (Editor in chief), 8718–21. 2nd edition. Farmington Hills: Thomson Gale, 2005.

Molesky-Poz, Jean. *Contemporary Maya Spirituality: The Ancient Ways Are Not Lost.* Austin: University of Texas Press, 2006.

Monod Becquelin, Aurore, and Cédric Becquey. "De las unidades paralelísticas en las tradiciones orales Mayas", *Estudios de Cultura Maya* XXXII (2008): 111–53.

Montes de Oca, Mercedes. "Los difrasismos: ¿núcleos conceptuales mesoamericanos?" In *La Metáfora en Mesoamérica*, edited by Mercedes Montes de Oca, 225–51. México: Universidad Nacional Autónoma de México, 2004.

Morales Sic, José Roberto. "Religión y Política: El proceso de institucionalización de la espiritualidad en el Movimiento Maya Guatemalteco". MA thesis. Guatemala: Facultad Latinoamericana de Ciencias Sociales, Guatemala, 2004.

Olupona, Jacob. *African Spirituality: Forms, Meanings, and Expressions.* New York: Crossroads, 2000.

Omoniyi, Tope, and Joshua A. Fishman. *Explorations in the Sociology of Language and Religion.* Amsterdam: J. Benjamins, 2006.

Ponciano, Karen. "Mayanización y experiencia religiosa: una lectura a partir de los aportes etnográficos sobre la espiritualidad maya". In *Mayanización y vida cotidiana. La ideología multicultural en la sociedad guatemalteca.* Volume 3, coordinated by Santiago Bastos and Aura Cumes, 283–305. Guatemala: FLACSO – CIRMA – Cholsamaj, 2007.

Richards, Julia Becker. "Case Study One: San Marcos La Laguna". In *The Life of Our Language: Kaqchikel Maya Maintenance, Shift, and Revitalization*, edited by Susan Garzon, R. McKenna Brown, Julie Becker Richards, and Wuqu' Ajpub, 62–101. Austin: University of Texas Press, 1998.

Rodríguez, Rosa María. "Usos sincréticos en el lenguaje ritual de costumbristas y pentecostales Mayas – K'iche'". Memorias del Congreso de Idiomas Indígenas de Latinoamérica II, University of Texas at Austin, 27–9 October 2005. Accessed 1 May 2013. www.ailla. utexas.org/site/cilla2/Rodriguez_CILLA2_k7iche7.pdf

Rojas Lima, Flavio. *Los Indios de Guatemala*. Madrid: MAPFRE, 1992.

Sheldrake, Philip. *A Brief History of Spirituality*. Malden, MA: Blackwell, 2007.

Sherzer, Joel. *Verbal Art in San Blas: Kuna Culture Through Its Discourse*. Albuquerque: University of New Mexico Press, 1990.

Tixal Mul, Israel, Pascuala Morales Calel, and Miguel Angel Vicente Reynoso. 2001. *Fuentes y fundamentos del derecho de la nación maya k'iche'*. Guatemala: Conferencia Nacional de Ministros de la Espiritualidad Maya, Oxlajuj Ajpop.

Voss, Alexander. "Continuidad sin cambio: fetichismo en la cultura Maya", *Ketzalcalli* 2007, no. 2 (2007): 49–62.

Watanabe, John. "From Saints to Shibboleths: Image, Structure, and Identity in Maya Religious Syncretism", *American Ethnologist* 117 (1990): 131–50.

Ximénez, Fray Francisco. *Historia de la provincia de San Vicente de Chiapa y Guatemala*. Guatemala: Tipografía Nacional, 1929–30.

3

WORDS TAKEN FOR WONDERS

Conversion and religious authority amongst the Dalits of colonial Chhattisgarh

Chad M. Bauman

The beginning of Homi Bhabha's influential essay, "Signs Taken for Wonders", relates the story of an early Indian Christian catechist who, in 1817, came upon a group of people gathered in a grove outside of Delhi. He asked them who they were, and they replied, "We are poor and lowly, and we read and love this book . . . the book of God." The catechist was surprised to learn that it was indeed the Bible. Pointing to the name of Jesus, the catechist asked, "Who is that?" The people replied, "That is God!" After further inquiry, the catechist learned that they received the book at the Hurdwar (Haridwar?) fair from a *pandit* they called an "angel of God". Concluding that the people must have encountered a missionary, he told them that the book taught "the religion of the European Sahibs" and that these Europeans had printed it in the vernacular for the use of Indians. The people could not accept that such a fine text had come from a race of meat-eaters and would not believe the catechist, preferring to hold to the notion that it had been sent directly from God. The catechist finally concluded, "The ignorance and simplicity of many are very striking, never having heard of a printed book before; and its very appearance was to them miraculous."[1] I must admit to finding the rest of Bhabha's article baffling and cryptic, and yet I have always found this particular passage significant for the way it portrays those (as the catechist put it) "ignorant" and "simple" people who had, like the Satnamis I studied in colonial Chhattisgarh, India, taken words for wonders.

In the early nineteenth century, an illiterate Chamar from what is now the central Indian state of Chhattisgarh, Guru Ghasidas, returned from a pilgrimage to the famous Jagannath temple with a message for his followers. Ghasidas's message was similar to that of many Hindu reformers who had come before him. He told his followers, amongst other things, to abandon the worship of images, to avoid eating meat, to reject the use of Brahmans as religious functionaries, and to devote themselves entirely to the one and only formless (*nirgun*) deity, whom he called "Satnam" (i.e. the True Name). The guru's following grew, slowly at first, but by his death in 1850 he claimed a quarter of a million disciples, and by the century's end, nearly all members of the Chamar caste in Chhattisgarh had joined his *panth* (a community of followers) and had begun calling themselves *Satnamis*.[2]

In 1868, eighteen years after Ghasidas's death, the first Western missionaries in the region, sent by German evangelicals living in the United States, arrived in Raipur and later set up camp nearby, in Bishrampur. They were joined, in 1885, by Disciples of Christ missionaries working outwards from Bilaspur. Given the downtrodden state of the *Satnami* community, these missionaries came to expect that all *Satnamis* would soon become Christian. Only a small percentage of *Satnamis* ever did convert, but the great majority of today's Christians in this area can trace their ancestry to the Satnampanth and thus to the Chamar caste.

This chapter argues that both Satnamis and Christians perceived the written word to possess an enigmatic power. Amongst these communities, the written word functioned in two primary ways. First, Christians and Satnamis agreed that a sacred text was a mark of religious authenticity. The fact that Satnamis had no text they could call their own, therefore, was a source of some communal insecurity. Christians and Christian missionaries at times exploited that insecurity for their own gain, and some Satnamis converted because they believed that the association of Christianity with books, and in particular with the Bible, proved its superiority to the faith of the Satnamis. Second, the written word functioned as a sign of epistemological authority. Consequently, the ability to write and publish texts gave Christians an advantage in religious debates with Satnamis. In this context, then, texts were not merely lifeless pieces of paper, but sacred and potent religious objects.

Introduction: learning to think (and read) like a Christian

Though it is clear that Chhattisgarhis altered and transformed the faith presented to them by foreign missionaries and native evangelists, it is equally clear that those who became Christian underwent a significant cultural transformation. They did come, in various ways, to assimilate the cultural prejudices and presuppositions of their evangelical interlocutors, in particular their methods of determining truth.[3] Religious affiliation entails not only the acceptance of an ultimate perspective (i.e. a worldview) but also commitment to particular forms of authority.[4] Conversion to Christianity in Chhattisgarh therefore involved the assimilation of a new epistemology. Many, including those who eventually became Christian, internalized this epistemology in ways of which they themselves were perhaps not entirely aware.[5]

The missionaries were, after all, modern individuals, confident in the inevitability of "logic" and the ineluctable "rationality" of their own religion. With the Basel missionaries, who in 1899 contrasted the "fancies" and cognitive "rubbish" of Indian religions with the "plain and chaste truths of the unadulterated Gospel",[6] missionaries in Chhattisgarh did not consciously recognize the ways in which their "logic" and "rationality" were reticulated with their own culture's broader assumptions regarding authority and proof.

In Chhattisgarh, the carriers of Christianity did understand, however, if only in an unconscious fashion, that their religion made little sense according to indigenous criteria. Nevertheless, despite that inkling of incongruity, they attempted, with various degrees of "success", to replace those criteria with their own. It was not that they understood the socially constructed nature of plausibility and reasoning. Rather, as modern individuals, they believed that they were in possession of universal "reason", while most Indians were not. They attempted, accordingly, to overcome this obstacle to conversion by inculcating the kind of "reason" necessary to ensure the desired result.[7]

The following parable, recorded by an evangelical catechist about a convert to Christianity on the evangelical mission field, evocatively describes the process by which Chhattisgarhis came to accept the missionary epistemology:

> [Jawal Pershad told a story about] a man [who] found a white pebble, [and] thinking it to be a diamond went to a jeweler to sell it for 10,000 rupees[. The] jeweler thought that the man was deceived by somebody and [that] if he would say that its value was nothing [the man] would be disheartened and would not believe anyone, [or] he would think that the jeweler wanted to cheat him, so he asked the man that if he consented to work at his shop he would give him sufficient wages per mensem. . . . The jeweler taught [him] the distinction between a real diamond and a false one, this he learnt in a year or two, when he was able to distinguish a real diamond from a false one, then [the jeweler] asked him to see his diamond, on seeing it he at once knew that it was a false one and showed no sign of sorrow. Jawal Pershad said that he loved the false religion which appeared to him to be true, but when he went amongst Christians, they taught him to distinguish between a false religion and a true one, [and said,] when I learnt this then I gave up the false one and embraced the true one which is Christianity.[8]

Like the ill-informed prospector who is able to distinguish false diamonds from real ones only after learning the jeweller's criteria for doing so, Pershad came to determine the "truth" of Christianity only after accepting Christian criteria for distinguishing "false" from "true" religions.

Christian missionaries and evangelists taught Chhattisgarhis to distinguish "false" from "true" religions by a variety of methods, of which the most important was education. Disciples of Christian missionary Homer Gamboe wrote in 1918: "The training of natives in secular education is a great essentiality in heathen lands. It is necessary to supply the individual with the tools of thought with which he is capable of forming conceptions of Christian faith before he can understand that faith."[9] By the 1930s, more than four hundred students were enrolled in evangelical mission institutions, and two thousand attended Disciples of Christ mission schools.[10] Twenty per cent of *Satnami*-Christian children attended school on the evangelical mission, as compared to only 1.3 per cent of Satnami children.[11]

The association of Christianity with education and literacy was reinforced at every opportunity. The entryway of the Disciples church in Bilaspur, completed in 1901, contained a prominent teacher's desk on which was displayed an open Bible, and around which were places for learners to sit.[12] Missionaries understood the power of their educational institutions. For example, Disciples missionary Donald McGavran was known for saying that he considered the children of converts, who would be educated in mission schools and become the "true" Christians, more important than the converts themselves. McGavran's colleague, George E. Miller, wrote, "When villagers become Christians, the chief question is not what can be made of them, but what can be made of their children?"[13] These educational institutions imparted an epistemological lens through which both Satnami and *Satnami*-Christian students came to interpret their world, a "sacred canopy",[14] to use Peter Berger's famous phrase, and socialized them into Western ways of thinking and reasoning such that they would be "more sympathetic listeners to the Gospel."[15]

The most salient of all effects of the missionary educational programme was to significantly increase literacy in Chhattisgarhi and Hindi amongst both Christians and members of other communities. Convinced Protestants that they were, missionaries and evangelists amongst the Satnamis feared that illiteracy would prevent their converts from being exposed to "The Word" in the Bible, tracts, and hymnbooks.[16] Missionaries expended a large portion of their resources, both human and financial, on teaching people to read. They produced magazines for mass consumption, and they published books, tracts, and pamphlets, mostly in Hindi and Chhattisgarhi, at a feverish pace. Sales of mission publications on the evangelical mission rose from four thousand in 1932 to nearly twenty-eight thousand in 1936.[17] Even those Christians who could not read were required to memorize texts such as the Ten Commandments, the Lord's Prayer, the Twenty-Third Psalm, and the Apostles' Creed.[18]

The Christian effort to eradicate illiteracy produced tangible results. By 1943, 28 per cent of Christian women on the evangelical mission field could read and write.[19] At roughly the same time, evangelical M.P. Davis figured that 40 per cent of *Satnami*-Christians were literate (as opposed to 8 per cent of Chhattisgarhis and 13 per cent of Indians).[20] In 1952, an unnamed Disciples missionary reported that 93 per cent of adults in the Mungeli congregation, most of them with Satnami ancestry, were literate.[21] These percentages were even higher than the average amongst all Christians in Chhattisgarh.[22]

Particularly amongst those who had been Christian for some time, the "value" of education and literacy seemed obvious, and *Satnami*-Christians were known to go to great lengths to get their children educated. The change was most pronounced amongst women and girls. Satnami converts and Christian evangelists assimilated the missionary epistemology and accepted the importance of education.[23] While literacy rates remained deplorably low, literacy itself came to be universally valued.

Historically, *Satnamis* and other Dalit communities had been denied access to traditional forms of education, and they were even generally barred from learning, reading, or writing Sanskrit, the language of Hindu scriptures and learning. Upper-caste Hindus such as Shankara (c. 788–820), and like-minded nineteenth- and twentieth-century scholars were loath to educate members of communities whose low birth, they believed, was a mark of karmic retribution. The Dalits had no karmic qualification for Vedic learning. Egalitarian access to the sacred text was, in this context, the most radical element of Christian teaching, and it is therefore not surprising that Satnamis who became Christian should so earnestly embrace the possibility of education.

Many non-Christian Satnamis, however, either did not consider education useful or recognized (and wished to avoid) the epistemological effects of missionary education. It was rumoured, in fact, that Ghasidas had at one time forbidden education to his followers (but that he eventually withdrew the objection). An evangelical, formerly Brahman evangelist, Ramnath Bajpai, reported that in the early twentieth century, some Satnami leaders had told a government official who wished to encourage their education, "We do not require these black letters, as [God] has ordained us to mind our work with the stick called Tutari [a stick with which one drives bullocks]."[24] Even into the 1930s, Satnami gurus discouraged members of the Satnami community from attending mission-run schools.[25] Nevertheless, as I argue below, the Satnami suspicion of education had more to do with Satnamis' intuition about its epistemological effects than with a distaste for knowledge per se. And what seems absolutely clear is that even non-Christian Satnamis held the written word in high regard – but more on that below.

Religious authenticity and the aura of the written word

The authority of written texts, as above that of oral traditions, was one distinctive aspect of the missionary epistemology. In this regard, the missionaries were merely exhibiting a bias which had been around since the Enlightenment. David Hume, for example, wrote in the *Natural History of Religion*:

> An historical fact, while it passes by oral tradition from eyewitnesses and contemporaries, is disguised in every successive narration, and may at last retain but very small, if any, resemblance of the original truth, on which it was founded. The frail memories of men, their love of exaggeration, their supine carelessness; these principles, if not corrected by books and writing, soon pervert the account of historical events; where argument or reasoning has little or no place, nor can ever recall the truth, which has once escaped those narrations.[26]

For the *Satnami*-Christian community, the Bible, a written, stable (by this era) text, became an evocative symbol of distinction and status. Attempting to convince Satnamis to join their community, Christians frequently contrasted their own literacy and access to a holy book with the illiteracy of Ghasidas, the "booklessness" of the Satnampanth, and the Satnamis' lack of access to Sanskritic Hindu scriptures.[27] The Satnamis, for their part, claimed (and still claim) the *Ramcharitmanas* as "their" book, and read other vernacular-language Hindu texts as well, though many were uncomfortably aware that these books reflected neither the perspective of their guru nor that of their *jati*. The facts that Ghasidas had not written a book, that the scriptures the *panth* claimed as its own contained no mention of Satnam, and that the community was denied access to the scriptures of Sanskritic Hinduism made the community vulnerable, as will be discussed below, to manipulation by Christian authors who, in written, *published* texts claimed to know the "true" history of Guru Ghasidas and to possess the "book of Satnam" (i.e. the Bible), the coming of which Ghasidas had allegedly predicted.[28] Christians played upon Satnami insecurities regarding the community's general illiteracy and lack of a sacred text. An unnamed author (presumably Indian) argued, for example, that "The Satnamis of

Chhattisgarh must be followers of a *pothi* (book). That *pothi* must be the *pothi* of Satnam, which means it must be the Bible."[29]

Tularam, who was born a Satnami but educated at mission schools, later chose Christianity in part because, unlike the Satnampanth, the community possessed a *book* that contained its main doctrines and tenets. According to one Satnami informant, Tularam

> called all the big people in the [Satnami] community and in the panth and in his family, with a Christian *padri* and Dr. [Donald A.] McGavran [to meet with him]. He said, "I am a Satnami, but I am going to convert; I am going to become a Christian. If you oppose me, then show me some religious book, some holy book of the Satnamis. If any Satnami book is present, show it to me, and I will not become a Christian. But because there is a book I am becoming Christian.[30]

Later, in the role of Christian evangelist, Tularam often emphasized this distinction between Christians and Satnamis as a potent symbol of the former's superiority.[31]

Given the fact that Satnamis had traditionally been denied access to Hindu religious texts, it is perhaps not surprising that the *Satnami*-Christian community's reverence for the Bible was perceived, at times, to border on idolization. Disciples missionary Donald McGavran was once greeted by a villager saying that there were in the area a few other people who, like him, "worshiped the book".[32] Though few *Satnami*-Christians appear to have considered the Bible an image to be worshipped, it did become something of a symbol of their superiority and status as a literate community. Bibles were valued gifts and prized possessions, even (or perhaps especially) amongst illiterate members of the community.[33] The same can be said of Satnamis and other non-Christians in Chhattisgarh, who purchased evangelical texts and tracts at a rate far higher than their own literacy or their interest in converting. The aura that attached to these texts was quasi-talismanic, and for *Satnami*-Christians they were potent symbols of newfound status and authority.[34] Non-Christian Dalits also revered texts, particularly Sanskritic texts, but from a distance and *because of their lack of access* to them. *Satnami*-Christians, on the other hand, valued the Bible as an unmistakable symbol of the fact that they were not only literate, but also encouraged to read the religious texts of their new community. Like education itself, reading and access to sacred texts symbolized one aspect of what they considered to be an appealing social vision.

Non-Christians understood that the reading of texts, religious or otherwise, somehow affected the reader in fundamental ways. In the 1940s, an evangelical missionary reported that some non-Christian leaders had instructed their followers not to purchase Christian literature because they believed it "did something to them".[35] The very act of reading, of being in intimate contact with the aura of the written word, was a momentous one imbued with the power to change people's minds and – more worrisome to these non-Christian leaders – one's religious affiliation.

Epistemological authority

From the very beginning, missionaries in Chhattisgarh attempted to forge connections between their faith and that of the Satnamis. They did this in a variety of ways, but the most important was their attempt to portray Christianity as the culmination and fulfilment of Guru Ghasidas's message.[36] The portrayal required a selective editing and rewriting of Satnami history and the conscious and unconscious re-mythologization of Ghasidas. Christians forged this link between the Satnampanth and Christianity in two ways: first, by emphasizing Ghasidas's alleged prophecy regarding the coming of a white-faced man and, second, by claiming that on his pilgrimage to Puri, Ghasidas had come into contact with Christian missionaries.

Early in his career, the first missionary in Chhattisgarh, German-American evangelical Oscar Lohr heard that a Satnami guru had prophesied the coming of a white man. The story, from an evangelical perspective, was almost too good to be true, and in part because of it missionaries in Chhattisgarh have been obsessed with converting the Satnamis ever since. Lohr sent home reports predicting that Christianity would spread quickly through familial ties and that very soon the "whole tribe" would "embrace Christianity".[37] Inspired by Lohr's glowing missives, a number of energetic young evangelical men set off for India to join the work, only to return again after a short period, disillusioned by the slow pace of conversions.[38]

In one of his first reports from Chhattisgarh, which was published in *Der deutsche Missionsfreund*, Lohr wrote (in German):

> Here I have to point out the reason of my favorable reception. The late guru (teacher) of the Satnamis, who was the third descendent of the founder of the sect, had at a certain occasion given the promise to his disciples, in a kind of prophetic excitement when there were ten thousand of them together, that their salvation would be brought to them by a sahib, or a white man and that they would be guided by this person on the right way, and that they should join with him with hope and trust.[39]

Later, in an autobiography, Lohr declared, "The Chamars accepted me as the teacher whose coming their leader had predicted, who would proclaim to them the way of salvation in a better way than he himself could do."[40] It seems clear that Lohr had intentionally positioned himself as this prophesied one – could he have resisted the temptation to point out that he was a white man with a message similar, in some ways, to that of Ghasidas?

Whereas Lohr's version of the account attributes the prophecy to the third Satnami guru, later versions of the story suggest that Ghasidas himself had uttered the prediction.[41] I trace the development of the Christian version of the story elsewhere,[42] but suffice it here to say that the Christian version of the Ghasidas legend eventually developed far more specific details, such as the fact that the coming white man would come from a Western country, would be red from sunburn, would wear boots and a hat, would have an office, would not use water to clean himself after using the toilet, and – most importantly – would be carrying a book under his arm.[43] These accounts, particularly the later ones, appear to combine Satnami legends about Ghasidas's prophetic abilities with the creative embellishment of Christians and missionaries who wished to claim the prophet's mantle, along with the allegiance of his followers.

There is no evidence from Satnami sources that such a prophecy ever occurred. Only older Satnamis educated by missionaries seem to have any knowledge of it. For example, one Satnami informant, Govind Das Patre, reported that Ghasidas had prophesied the arrival of white-faced people with golden hair who would come from the East (the closest area of missionary and British strength was in and around Cuttack and Puri, to the East) with books under their arms, preaching about "*satya*". When asked where he had heard the story, he said, "The *padris* [missionaries] told it like that." He later insisted that both missionaries and older Satnamis told the story in this way.[44]

Whereas the Christian account of Ghasidas's prophecy may have been drawn from some "real", if insignificant, element of the Satnami tradition, Christian accounts of the guru's pilgrimage to Puri, and what happened along the way, appear to be largely imagined, the inventions of Satnami converts and their Christian leaders. But though they may be fictions, they are *plausible* fictions, and therein lies their power to evoke a past (i.e. the legacy of Ghasidas) and to invoke a present (i.e. continuity with Christian history). The most common traditional Satnami version of events suggests that Ghasidas encountered Satpurush ("true, primordial man"), a mysterious spiritual being, while on a pilgrimage to the Jagannath temple in Puri. Early British reports of Ghasidas's inspiration, such as that in Grant's 1870 *Gazetteer of the Central Provinces*, follow this account and mention no connection between Ghasidas and missionaries. This suggests that before the arrival of missionaries the *panth* itself did not believe there to be a connection between Ghasidas and Christianity.[45] Likewise, British ethnographic texts and gazetteers published well into the twentieth century follow this traditional version of the story.[46] So do contemporary Satnami accounts.[47]

Christian recensions, which assert that Ghasidas acquired his message from Christian missionaries whom he encountered on his pilgrimage to Puri, first emerged around the turn of the century. A pamphlet written by Mary Kingsbury is typical and amongst the earliest archival records of the emerging Christian version of the story:

> There is a tradition that [a *Chamar*] went on a pilgrimage to the famous shrine of Juggernaut, and while there he heard a white *Guru*, or teacher, who spoke about the sin of idolatry, saying there was only one true God, and that He alone should be worshiped. When this man, who was named Ghussee Das, returned to his country, he determined to establish this new religion among his people. In order to do this, he separated himself from them for a while, going into the jungle to live, and telling his people that God would give him a message, and he would come again and deliver it to them . . . When, after about six months, he did appear, the people received him as a prophet. . . . You will judge aright from this account, that Ghussee Das must have listened to a missionary preaching when he was at Juggernaut, and that although many things were confused in his mind, he had received a little of the truth which remained with him and bore some fruit.[48]

In 1909, Disciples missionary E.M. Gordon echoed Kingsbury's account, asserting that it was "more than likely that during his stay in Jagganath [Ghasidas] heard the preaching of the early missionaries."[49]

Like the story of Ghasidas's prophecy, the Christian account of the guru's encounter with missionaries accumulated greater detail over the following decades. Already by 1897, some evangelical missionaries were asserting that the encounter between Ghasidas and missionaries occurred in Cuttack (not Jagganath).[50] In 1934, H.C. Saum wrote to a certain Reverend Lazarus, pastor of the Baptist church at Cuttack, asking whether Lazarus could verify the "tradition" regarding Ghasidas's visit to the city. No record exists of a reply from Lazarus, but Saum's letter indicates that by that time another detail had been added to the traditional Christian version of the story – that the missionaries were Baptist (no other missionaries were active around Cuttack at that time).[51] In a book published in 1990, but based primarily on his experiences in the 1930s and '40s, Disciples missionary Donald McGavran, an inveterate mythologizer, suggested that Ghasidas might have, in fact, met William Carey himself – a possibility, of course, but a long shot based on little more than McGavran's own desire for it to be so.[52]

The reinterpreted, Christianized Ghasidas provided the *Satnami*-Christian community with a link not only to its past, but also to the larger Christian story. The guru was portrayed as both a conduit of the Christian message and as a precursor of Christian missionaries. According to the Christian version of events, conversion to Christianity not only became the logical choice for those wishing to abide by Ghasidas's quasi-Christian message, but also represented a return to the newly discovered Christian roots of his inspiration. Rather than abandoning their tradition, Satnami converts to Christianity were returning to the mythic past.

Consciously or unconsciously, leaders of the *Satnami*-Christian community began to fashion Ghasidas in the image of John the Baptist. Ghasidas made it easy for them; he fit the bill. Like John the Baptist, he was an eccentric character, a bit wild and woolly; he preferred to spend his time in the wilderness, and his unrefined and unlettered nature contrasted dramatically with the literate purveyors of the dominant religious orthodoxy. According to a missionary author, Ghasidas said:

> a day will come when the True Name will make his people pure. When my elder brother comes (or one *greater* than I) he will teach the old Book from house to house and in that day there will be drinking from the same pool the lion and the cow (– meaning that they will not devour each other as does the caste system). I, the prophet, am not the true name, nor do I have the old book of the True Name. I am not able to give you the entire secret, but he who will come in the future, will reveal the true God to you. . . . A WHITE MAN WEARING A HEAD COVERING (helmet). HE WILL HAVE THE OLD BOOK IN HIS HAND, AND WILL TELL YOU THE SECRET OF THE TRUE BOOK [emphasis in the original].[53]

In this striking mélange of religious mythologies, the missionary puts into Ghasidas's mouth the words of John the Baptist, "There cometh one mightier than I after me", (Mark 1:7 KJV, see also Matt. 3:11 and Luke 3:16), and places him squarely within the Hebrew and Christian prophetic tradition by having him utter a localized version of Isaiah's prophecy: "The wolf also shall dwell with the lamb, and the leopard shall lie down with the kid; and the calf and the young lion and the fatling together; and a little child shall lead them" (11:6 KJV). This missionary was not alone in portraying Ghasidas as a Christian precursor and Christian missionaries as the fulfilment of his prophecies.[54] A tract written around 1920 by evangelical catechist and pastor Ramnath Bajpai (who added "Simon" to his name when he was baptized), says:

> As Ghasidas told you that white sahib would come and would bring along with him the name of [Truth] or True name, [so you should] open your eyes and see that the white faced sahabs [*sic*] have come, and they really preach about Jesus Christ the True name . . . Get hold of him and come out of the mud. Ghasidas told you that when those white sahebs [*sic*] come there will be the dawn of [a] True age with the effect that tigers and cows would drink water from the same drinking place or ghat [a stone stairway down to water] together. Is it not true? Yes, it is true, Brahmins and the Chamars drink water from the same pipe.[55]

After forming Ghasidas in the image of John the Baptist, leaders of the *Satnami*-Christian community went on to assert that the Satnami community had fallen from the standards set by its guru. The source of Ghasidas's message, after all, had been pure, and whatever slight differences there may have been between Christianity and the message preached by Ghasidas were attributed to the brevity of his exposure to Christianity, his illiteracy, or his simplicity. It was his successors, especially Balakdas, who debased a movement Christian leaders regarded as positive overall, and who had dragged the Satnami community down once again into polytheism and – referring to an occasionally performed fertility rite called *satlok* which involved ritualized sexual intercourse – immorality.[56] Though the scabrousness of the Satnamis' alleged sexual profligacy provided an evocative foil on which the Christian community's relative chastity and sexual fidelity appeared all the more striking, it was the movement of Satnamis away from their founder's rejection of image worship that most provoked the consternation of Chhattisgarhi Christians and missionaries.

After the 1920s, while Satnami leaders were attempting rapprochement with upper-caste Hindu leaders and were Sanskritising their rituals and beliefs, Christians attempted to convince them that they were *not* Hindus[57] and reminded them that Hindus had formerly considered them "outside the fold." As Christians presented the choice, Satnamis could either embrace Ghasidas's prohibitions against image-worship by becoming Christian or reject Ghasidas's instructions in order to seek the approbation of oppressive upper-caste Hindus. By asserting that the

Satnampanth had fallen away from the unspoiled message of its founder, Christians were able to portray conversion to Christianity not as a rejection of the Satnampanth, but, as one *Satnami*-Christian put it, a return "*back* [emphasis added] to Satnam."[58] Joining the Christian community was about rediscovering and embracing the original, but now vitiated, message of Ghasidas, about digging through the detritus of a corrupt religion and returning to its unspoiled source.

As the preceding discussion suggests, leaders of the Disciples' Christian community appear to have been far more interested in the history of Ghasidas than their evangelical coreligionists. Nevertheless, in no single text is the Christian reinvention of Ghasidas and the Satnampanth more clearly, more comprehensively, or more coherently laid out than in *Satyanami Panth: Shri Gosain Ghasidas of Girod*[59] (1937), written by M.M. Paul, an evangelical catechist and pastor working (at the time) in Pithora.[60] But it was not the first text of its kind. I have already mentioned Simon Ramnath Bajpai's earlier text, *Some Facts about Satnamis* (ca. 1921). Paul's text shares much in common with Bajpai's, and at times the two are nearly exact replicas, from the fundamental arguments they make, to the account and interpretation of Ghasidas's story, to the quirky idiomatic expressions they employ (such as their admonition that Satnamis should "get out of the mud" and follow the true Satnam). Paul's claim to have heard his version of the story from an early Satnami convert notwithstanding (Bajpai does not make this claim), it is clear either that he was familiar with Bajpai's earlier tract or that both drew upon an as-yet undiscovered *Ur*-text, whether a written work or a fairly common oral account. Bajpai and Paul both worked for the evangelical mission, and it is therefore entirely plausible that Bajpai's work was passed along to Paul for updating and publication.

Though it is nearly impossible to ascertain how many, and what kind of people read the tract, M.M. Paul's *Satyanami Panth* established the standard Christian version of the Satnami story, and most subsequent accounts of the events, on both the Disciples and evangelical mission fields, follow the outlines of his narrative. The text drew authority from the fact that it was *written*; it was a tangible, durable, stable document. The importance of this fact, in a context of low literacy but high regard for literate specialists, cannot easily be overstated.

In fact, *Satyanami Panth* and Bajpai's *Some Facts about Satnamis* contributed to the standardization of the previously diverse and more fluid oral tradition from which they had been drawn. During an interview with Christians from Ganeshpur, all accounts of the story of Ghasidas roughly followed the shape of Paul's version of events, but the informants were divided on whether they had heard the story from elders or read it in "some book".[61]

Similarly, during his field research, Saurabh Dube encountered a Christian whose oral history of Ghasidas also closely followed Paul's account. When asked where he had heard it, the man appeared upset and said, "It is the true thing", adding that it was *likhit itihas* (written history).[62]

Even Satnamis, particularly those who had contact with Christians or were educated in their schools, were influenced by this written "history" of their *panth*. In the 1950s, Satnami leaders were still concerned about the effects of *Satyanami Panth* on the Satnami community. Naindas and Anjordas, leaders of the Satnami Mahasabha, both testified about the text before the Christian Missionary Activities Enquiry Committee, also known as the Niyogi Commission, which investigated claims of induced conversion to Christianity for the Madhya Pradesh state government. Anjordas told the committee that missionaries had been saying that Ghasidas met missionaries at Cuttack and that Ghasidas had become a Christian. Anjordas then referred to Paul's *Satyanami Panth* as evidence of missionary wrongdoing, saying:

> A book [alleging that Ghasidas had been inspired by Christian missionaries] has been published and widely circulated amongst the Satnamis. On such type of preaching many Satnamis have become Christians. . . . To my knowledge, about a thousand Satnamis became converts after the story was circulated. I do not know how many Satnamis actually read the book containing the story of Ghasidas. Many Satnamis who are in possession of the book . . . have not read it.[63]

Clearly, Paul's text and the version of events it contained influenced far more Satnamis than merely those who had become Christian.

Conclusion

Though *Satnami*-Christians had assimilated a certain reverence for reading and the written word from modern missionaries, it is clear that long before the arrival of European Christians there had been in Chhattisgarh, as elsewhere in India, a veneration for writing, for words, and for those who could interpret them. The fact that the Dalit Satnamis were excluded from the world of words served only to make those words all the more mysterious and, therefore, potent symbols of status and authority. Some Satnamis, therefore, converted to Christianity in part because Christians were encouraged to read and become literate. Education and literacy did, of course, have economic advantages, but more importantly, access to words, and particularly *sacred* words reflected an egalitarian social vision that many Satnamis found attractive. Moreover, the mere fact that Christians *possessed* a sacred book, unlike the Satnamis, who had no book they could truly and convincingly call their own, was enough, as shown above, to persuade some Satnamis of the superiority of the Christian faith. Words, therefore, served as a marker not only of status, but also of religious authenticity.

But words, especially the ability to *print* words, also allowed Christians, both missionary and Indian, to reinvent and reinterpret the story of Ghasidas in such a way that conversion to Christianity seemed like a fulfilment of Ghasidas's vision, rather than a rejection of it. Because it was printed, this reinterpreted story supplanted earlier, probably more accurate oral accounts of Ghasidas's life. Printed words, in this case, thereby became a sign of epistemological authority. As early as 1908, some non-Christian Satnamis had developed a modicum of scepticism towards the texts distributed by their Christian neighbours. Some villagers responded to a catechist's claim that he had books that would disprove their religion by saying, "these books have been printed by you, you may insert whatever you like."[64] Nevertheless, even many non-Christian Satnamis, like the Govind Das Patre (quoted above), came to accept the Christian account as the "true" one. In this context, therefore, words – printed words – were more than ink on paper. They possessed a certain power, undefined and yet undoubtedly real; they were words taken for wonders.

Notes

1 Homi K. Bhabha, "Signs Taken for Wonders", in *The Post-Colonial Studies Reader*, edited by Bill Ashcroft, Gareth Griffiths, and Helen Tiffin (London: Routledge, 1995): 29–30. Sections of this chapter first appeared in Chad M. Bauman, *Christian Identity and Dalit Religion in Hindu India, 1868–1947* (Grand Rapids, MI: Wm. B. Eerdmans), though the chapter represents a significant refocusing/reframing of the argument/s made in that context.

2 In this paper, "*Satnami*" (in italics) refers always to the caste, and "Satnami" (no italics) refers to a follower of Ghasidas. Saurabh Dube, *Untouchable Pasts: Religion, Identity, and Power Among a Central Indian Community, 1780–1950* (Albany: State University of New York Press, 1998): 50; India Census Commissioner, *Census of India, 1901*, Volume 8 (Part 2, Statistics) (Bombay: Government Central Press, 1902–3).

3 Cf. David A. Snow and Richard Machalek, "The Sociology of Conversion", *Annual Review of Sociology* 10 (1983): 170 ff.

4 Charles F. Keyes, "Why the Thai Are Not Christians: Buddhist and Christian Conversion in Thailand", in *Conversion to Christianity: Historical and Anthropological Perspectives on a Great Transformation*, edited by Robert W. Hefner (Berkeley: University of California Press, 1993): 262.

5 Jean Comaroff makes the same claim regarding the Tshidi in South Africa. Jean Comaroff, *Body of Power, Spirit of Resistance: The Culture and History of a South African People* (Chicago: University of Chicago Press, 1985).

6 Mrinalini Sebastian, "Reading Archives from a Postcolonial Perspective", *Journal of Feminist Studies in Religion* 19, no. 1 (2003): 22.

7 The assertion is, of course, not new. Evans-Pritchard claimed as much for the missionaries he observed: "Much of what [missionaries] teach natives is quite unintelligible to those among whom they labour, and many of them would, I think, recognize this. The solution often adopted is to transform the minds of native children into European minds." E.E. Evans-Pritchard, *Theories of Primitive Religion* (Oxford: Clarendon Press, 1965): 14. American missionaries in Chhattisgarh were not the first missionaries in India to believe in the superior rationality of their faith. Francis Clooney asserts that Roberto de Nobili, the early-seventeenth-century Jesuit missionary to India, believed that by "appropriating the truths accessible to reason, humans remove obstacles which prevent them from being ready to accept the higher truths of Christianity; a truly reasonable person can have no objection to Christianity." Francis X. Clooney, S.J., "Roberto De Nobili, Adaptation and the Reasonable Interpretation of Religion", *Missiology* 18 (1990): 32. What makes de Nobili so important in this connection is that in so many other respects he accommodated himself to (Brahmanical) cultural norms.

8 Entry in diary, unnamed catechist, 16 February 1909, found in the journal of S.J. Scott, Mahasamund catechist, Archives of the Evangelical Synod at Eden Theological Seminary, St Louis, Missouri (hereafter AES): 83–5 Di54 #1.

9 Homer P. Gamboe, "The Missionary Work of the Disciples of Christ in India and Its Development" (BDiv thesis, The College of the Bible, 1918): 72.

10 On the evangelicals, see J.W. Pickett, *Church Growth and Group Conversion* (Lucknow: Lucknow Publishing House, 1956). For Disciples mission stats, see Donald A. McGavran, "Character-Building En Masse as Conducted in India", *World Call*, March 1931, 28.

11 Davis, M.P., "A Study of Christians [*sic*] Descendants from Chamars and Satnamis", 1943, AES 80–1 Sat8.

12 Gamboe drew the distinction between Christian churches, which encouraged education, and Hindu temples, which did not: "No such idea prevails in temples. They are without Bible, rostrum, or auditorium. Teaching is not in their plans." Gamboe, "The Missionary Work", 26.

13 George E. Miller, "The Serpent That Failed", *World Call*, July 1922, 28. More information on McGavran was given by Yishu Prasad, Takhatpur, CG, 19 February 2004; Rajendra and Chandra Kala Lal, Jerhagaon, CG, 19 February 2004; and Elnora Scott, Barella, CG, 30 March 2004 (interviews by author).

14 Peter Berger, *The Sacred Canopy: Elements of a Sociological Theory of Religion* (Garden City, NJ: Doubleday, 1967).

15 Evangelical missionary "Sadhu" Hagenstein, quoted in Martin P. Davis, *Sadhu Hagenstein: A White Man among the Brown* (St Louis: The Board for Foreign Missions, Evangelical Synod of North America, 1930): 140.

16 Emil W. Menzel, *I Will Build My Church: The Story of Our India Mission and How It Became a Church* (Philadelphia: Board of Missions, Evangelical and Reformed Church, 1943): 32.

17 See P.D. Gottlieb, "Annual Report for 1931–32", AES 82–13a Pr91, Annual Reports, no. 17, 1932; and "Report of the President, 1935–36", ibid., no. 21, 1936.

18 Donald A. McGavran, *The Satnami Story: A Thrilling Drama of Religious Change* (Pasadena, CA: William Carey Library, 1990): 41.

19 Menzel, "I Will Build My Church", 69.

20 Martin P. Davis, *India Today and the Church Tomorrow* (Philadelphia: The Board of International Missions, Evangelical and Reformed Church, 1947): 59–60.

21 "Brief Report on the Christian Church, Mungeli", circa 1952, Disciples of Christ Historical Society (hereafter DCHS), United Christian Missionary Society Records, Division of Overseas Mission, India, box 14 (folder: India-Orientation). These and earlier figures do not distinguish between those who were literate in Hindi and those who were literate in Chhattisgarhi.

22 Christianization did not always entail great increases in literacy, and due to mass movements in other parts of India in the 1920s, which brought large numbers of illiterate Christians into the fold and overwhelmed missionary institutions, Christians, who had previously ranked first in literacy, fell to fourth (after the Parsis, Jains, and Buddhists). See "Official Census of India (Reprinted from the Missionary Herald)", *World Call*, July 1924, 37.

23 On Nehemiah Goreh, a prominent nineteenth-century Indian Christian convert whose "transition from Hinduism was affected by his perception of how Christians think as much as what they think," see Richard Fox Young, "Some Hindu Perspectives on Christian Missionaries in the Indic World of the Mid Nineteenth Century", in *Christians, Cultural Interactions, and India's Religious Traditions*, edited by Judith M. Brown and Robert Eric Frykenberg (Grand Rapids, MI: William B. Eerdmans, 2002): 54–5.

24 *Some Facts about the Satnamis*, 1921(?), DCHS, H.C. Saum Files, box 2 of 4 (no folder).

25 Martin P. Davis, "The Satnami Tragedy," *Evangelical Herald*, 1 June 1933, 429.

26 David Hume, *Writings on Religion*, edited by Antony Flew (La Salle, IL: Open Court, 1992): 112.

27 See, for example, H.C. Saum, "Who Are the Satnami?" 1936(?), DCHS, H.C. Saum Files, box 2 of 4 (folder: Dr. Don Sat Mss.); and M.M. Paul to F.A. Goetsch, 29 November 1937, AES 89–2c3 Dev49, carton 1, #3. Evangelists could not distinguish Christians from Hindus in the same way, as Hindus could legitimately and without reservation claim the Vedas as their own. When preaching to Hindus or Kabirpanthis (who also possessed their own scriptures), at least one unnamed evangelical catechist argued for the superiority of Christianity based on the more manageable size and language of its scriptures. See his journal, entries for 26 June and 6 November 1908, found in the journal of S.J. Scott, Mahasamund catechist, AES 83–5 Di54 #1.

28 Christian authors claimed not only that Ghasidas had prophesied that missionaries would come and bring "the book of Satnam", but also that because of this book Brahman scriptures would be destroyed and "eaten up by worms". See Ramnath Bajpai, *Some Facts about the Satnamis*, circa 1921, HCS, box 2 of 4 (no folder). Similarly, M.M. Paul wrote that Ghasidas predicted that because of the missionary Bible, the knowledge of the Brahmans

would "become useless" and their books would "become meaningless". M.M. Paul, *Satyanami Panth: Shri Gosain Ghasidas Girod Vasi* (Allahabad: The Mission Press, 1937): 7, 21.

29 The author went on to say that "The Satnamis have no scriptures [*shastra*], no definite tenets [*nishchit bat*], and no proper statements." *"Satnam panth aur satnam ki pothi* [The Satnampanth and the Book of Satnam]" undated, found in HCS, box 1 of 4 (folder: Satnami Mss).

30 Govind Das Patre, interview by author, Takhatpur, CG, 23 March 2004.

31 Jayant R. Paul, interview by author, Takhatpur, CG, 27 March 2004.

32 Donald A. McGavran, "The Spread of the Word in India", 1939, HCS, box 2 of 4 (folder: Personal Letters and Memories).

33 Frances Waller Gamboe, "How They Did It [Pentecost] in Mungeli", *World Call*, June 1930, 26.

34 Cf. Dick Kooiman, *Conversion and Social Equality in India* (New Delhi: South Asia Publications, 1989): 190.

35 Davis, *India Today and the Church Tomorrow*, 19.

36 Christian leaders in Chhattisgarh thus placed Christianity in the same position in relation to the Satnampanth as Farquhar did in relation to all of Hinduism, that is, as its *fulfilment*; see J.N. Farquhar, *The Crown of Hinduism* (London: Oxford University Press, 1913). On Farquhar, see Eric Sharpe, *Not to Destroy But to Fulfil: The Contribution of J.N. Farquhar to Protestant Missionary Thought in India Before 1914* (Uppsala, Sweden: Almqvist & Wiksells Boktryckeri AB, 1965).

37 "Third Annual Report of the Chutteesgurh Mission, 1871–72", AES 82–13a Pr91, 1870–93 (Annual Reports). See also Oscar Lohr, "Von Darchoora", *Der deutsche Missionsfreund* 4, no. 4 (1869): 1.

38 Theodore C. Seybold, *God's Guiding Hand: A History of the Central Indian Mission 1868–1967* (Philadelphia: United Church Board for World Ministries of the United Church of Christ, 1971): 26–7.

39 Oscar Lohr, "Leiden Und Freuden Unseres Missionars," *Der deutsche Missionsfreund* 4, no. 1 (1869): 5.

40 "The Autobiography of the Rev. Oscar T. Lohr, in 1900", AES 84–5 Lo78.

41 To my knowledge, the earliest assertion that Ghasidas uttered the prophecy came in 1894; see Th. Tanner, *Im Lande Der Hindus Oder, Kulturschilderungen Aus Indien* (St Louis: n.p., 1894): 35.

42 Chad M. Bauman, "Identifying the Satnam: Hindu Satnamis, Indian Christians, and Dalit Religion in Colonial Chhattisgarh, 1868–1947" (PhD Dissertation, Princeton Theological Seminary, 2004): 200 ff.

43 See, for example, E.M. Gordon, *Indian Folk Tales: Being Side-Lights on Village Life in Bilaspore, Central Provinces* (London: Elliot Stock, 1909): 5; Bessie Farrar Madsen, "Come and Teach My People: A Call from India", *Missionary Tidings*, June 1914, 57; H.C. Saum and Mildred Saum, "Followers of the True Name", *Mungeli News Letter*, March 1935, 3.

44 Govind Das Patre, interview by author, Takhatpur, CG, 23 February 2004.

45 Charles Grant, editor, *The Gazetteer of the Central Provinces of India* (New Delhi: Usha, 1984). See also *Notes on the Sutnamee Chumars of the Raepore District from Information Gathered Orally from the Gooroo, Agur Dass, and His Disciples, and from Chumars Generally in 1868*, unsigned supplement to Temple's 1867 *Gazetteer of Central India*, DCHS, H.C. Saum files, box 3 of 4 (no folder).

46 See, for example, A.E. Nelson, *Central Provinces District Gazetteer: Raipur District* (Bombay: British India Press, 1909), Robert Vane Russell, *The Tribes and Castes of the Central Provinces of India*, Volume 1 (London: Palgrave Macmillan, 1916).

47 Govind Das Patre, Takhatpur, CG, 23 February 2004; and Sukh Chand, Pyarelal (informant uses only one name), Premsai Kossle and Awadh Ram, Kuran Kapa, CG, 29 April 2004 (interviews by author).

48 Mary Kingsbury, *Bilaspur* (Indianapolis: Christian Women's Board of Missions, n.d.): 5. The text, unfortunately, is undated, and Kingsbury worked nearly continuously in Bilaspur from 1885 until her retirement in 1925. But internal evidence, including the spelling of "Ghussee Das", suggests a publication date between 1895 and 1905.

49 Gordon, *Indian Folk Tales*, 4.

50 See Julius Richter, *History of Missions in India*, trans. Sydney H. Moore (New York: Revell, 1906): 251. Richter here quotes an unknown Evangelical mission publication of 1897.

51 H.C. Saum to Rev. Lazarus, 20 November 1934, DCHS, H. C. Saum Files, box 4 of 4 (folder: Satnamis). It is unlikely that Baptist missionaries would have remembered a visit from an unknown Satnami who had not yet been established as a powerful guru, so even a negative response would not necessarily have ruled out the possibility of an encounter with Ghasidas. For one text which identifies the missionaries as Baptist (but which gives no specific names), see Saum and Saum, "Followers of the True Name", 3.

52 McGavran, *Satnami Story*, 1, 12.

53 "The Christian Approach to the Satnami," undated (circa 1935), probably written by Donald A. McGavran, DCHS, United Christian Missionary Society Records, Division of Overseas Mission, India, box 41, McGavran, Donald A., Mr and Mrs (folder: India II – McGavran, Mr and Mrs Donald).

54 McGavran claimed that when he and others preached to Satnamis, they would say "Your own guru, Ghasi Das, foretold a hundred years ago that there would come a red-faced man with a big hat and a big book in his hand. Well, this is the red-faced man and this is the big book. Listen to what it says. It *amplifies* what Ghasi Das proclaimed." McGavran, *Satnami Story*, 42.

55 *Some Facts about the Satnamis*, 1921(?). The tract, which I discovered in the files of the Disciples missionary H.C. Saum, is a typescript English translation of an originally Hindi text. It is undated and unsigned. But in a letter from H.C. Saum to evangelical missionary Dr Jacob Gass (25 May 1935), Saum quotes from the tract and attributes it to "your man, Rev. Ramnath Bajpai". Seybold reports that Bajpai was ordained on 25 April 1920 and died on 16 December 1922; see Seybold, 1971. Therefore, the text could have been written no later than 1922 and, assuming Saum was correct in designating Bajpai a "Reverend", no earlier than 1920. Both the tract and the letter were found in DCHS, H.C. Saum Files, box 2 of 4; the tract was loose (no folder) and the letter was in a folder entitled "Bishrampur History Data".

56 See, for example, Simon Patros, *Simon Patros of India: A Miracle of God's Grace* (St Louis: Board of International Missions, Evangelical and Reformed Church, Inc., 1954): 43. For more on *satlok*, see Dube, *Untouchable Pasts*, 103–5.

57 One evangelical catechist, who complimented Satnamis on their monotheism, reported, "They take great pleasure to hear that they are higher than Hindus, because they have given up idolatry and worship the one true God". Entry in diary, unnamed catechist, 22 February 1908, found in the journal of S.J. Scott, Mahasamund catechist, AES 83–5 Di54 #1.

58 Hira Lal, *Satnam ke pas laut ao* (Come Back to Satnam, circa 1938), published tract (with no publication details), found in DCHS, H.C. Saum Files, box 2 of 4 (folder: Tularam). Missionaries also sounded the theme of fallenness. See Nelle Grant Alexander, *Disciples of Christ in India* (Indianapolis: United Christian Missionary Society, 1946): 9; Edith Moulton Melick, *The Evangelical Synod in India* (St Louis: Eden Publishing House, 1930): 25, and Menzel, *I Will Build My Church*, 18. British officials shared with missionaries an evolutionary sense of monotheism as the highest form of religion and therefore of polytheism as a corruption. Nelson asserted, "The crude myths which are now associated with the story Ghasi Das . . . furnish a good instance of the way in which a religion, originally of a high order of morality, will be rapidly degraded to their own level when adopted by a people who are incapable of living up to it." Nelson, *Central Provinces District Gazetteer*, 81. See also Grant, editor, *Gazetteer of the Central Provinces*, cxxx.

59 The Hindi text spells it "Giraud", but "Girod" is the spelling more common today.

60 "Shri" is a customary mark of respect, and "gosain" is an honourary title, the latter usually given to holy men. All translations of the text, which was written in Hindi and Chhattisgarhi, are mine. Special thanks to Dr Manish Tiwari for help with thorny passages.

61 Ganeshpur Christian group interview, Bishrampur, CG, 23 March 2004.

62 See Dube, *Untouchable Pasts*, 204.

63 Christian Missionary Activities Enquiry Committee, *Vindicated by Time: The Niyogi Committee Report on Christian Missionary Activities (1956)* (New Delhi: Voice of India, 1998): Volume 2, part B, 98. For other similar complaints, see ibid., 78–9.

64 Entry in unnamed catechist's journal for 26 September 1908, found in the journal of S.J. Scott, Mahasamund catechist, AES 83–5 Di54 #1.

References

Alexander, Nelle Grant. *Disciples of Christ in India*. Indianapolis: United Christian Missionary Society, 1946.

Bauman, Chad M. "Identifying the Satnam: Hindu Satnamis, Indian Christians, and Dalit Religion in Colonial Chhattisgarh, 1868–1947". PhD Dissertation, Princeton Theological Seminary, 2004.

Bauman, Chad M. *Christian Identity and Dalit Religion in Hindu India, 1868–1947*. Grand Rapids, MI: Wm. B. Eerdmans, 2008.

Berger, Peter. *The Sacred Canopy: Elements of a Sociological Theory of Religion*. Garden City, NJ: Doubleday, 1967.

Bhabha, Homi K. "Signs Taken for Wonders". In *The Post-Colonial Studies Reader*, edited by Bill Ashcroft, Gareth Griffiths and Helen Tiffin, 29–35. London: Routledge, 1995.

Christian Missionary Activities Enquiry Committee. *Vindicated by Time: The Niyogi Committee Report on Christian Missionary Activities (1956).* New Delhi: Voice of India, 1998.

Clooney, Francis X., S.J. "Roberto De Nobili, Adaptation and the Reasonable Interpretation of Religion". *Missiology* 18 (1990): 25–36.

Comaroff, Jean. *Body of Power, Spirit of Resistance: The Culture and History of a South African People.* Chicago: University of Chicago Press, 1985.

Davis, Martin P. *Sadhu Hagenstein: A White Man among the Brown.* St Louis: The Board for Foreign Missions, Evangelical Synod of North America, 1930.

Davis, Martin P. "The Satnami Tragedy", *Evangelical Herald*, 1 June 1933: 429.

Davis, Martin P. *India Today and the Church Tomorrow.* Philadelphia: The Board of International Missions, Evangelical and Reformed Church, 1947.

Dube, Saurabh. *Untouchable Pasts: Religion, Identity, and Power among a Central Indian Community, 1780–1950.* Albany: State University of New York Press, 1998.

Evans-Pritchard, E.E. *Theories of Primitive Religion.* Oxford: Clarendon Press, 1965.

Farquhar, J.N. *The Crown of Hinduism.* London: Oxford University Press, 1913.

Gamboe, Frances Waller. "How They Did It [Pentecost] in Mungeli", *World Call*, June 1930: 26.

Gamboe, Homer P. "The Missionary Work of the Disciples of Christ in India and Its Development". BDiv thesis, The College of the Bible, 1918.

Gordon, E.M. *Indian Folk Tales: Being Side-Lights on Village Life in Bilaspore, Central Provinces.* London: Elliot Stock, 1909.

Grant, Charles, editor. *The Gazetteer of the Central Provinces of India.* New Delhi: Usha, 1984.

Hume, David. *Writings on Religion.* Edited by Antony Flew. La Salle, IL: Open Court, 1992.

India Census Commissioner. *Census of India, 1901.* Volume 8 (Part 2, Statistics). Bombay: Government Central Press, 1902–3.

Keyes, Charles F. "Why the Thai Are Not Christians: Buddhist and Christian Conversion in Thailand". In *Conversion to Christianity: Historical and Anthropological Perspectives on a Great Transformation*, edited by Robert W. Hefner, 259–84. Berkeley: University of California Press, 1993.

Kingsbury, Mary. *Bilaspur.* Indianapolis: Christian Women's Board of Missions, n.d.

Kooiman, Dick. *Conversion and Social Equality in India.* New Delhi: South Asia Publications, 1989.

Lohr, Oscar. "Leiden und Freuden Unseres Missionars". *Der deutsche Missionsfreund* 4, no. 1 (1869): 5–6.

Lohr, Oscar. "Von Darchoora". *Der deutsche Missionsfreund* 4, no. 4 (1869): 1–2.

Madsen, Bessie Farrar. "Come and Teach My People: A Call from India". *Missionary Tidings*, June 1914: 57–8.

McGavran, Donald A. "Character-Building En Masse as Conducted in India", *World Call*, March 1931: 28.

McGavran, Donald A. *The Satnami Story: A Thrilling Drama of Religious Change*. Pasadena, CA: William Carey Library, 1990.

Melick, Edith Moulton. *The Evangelical Synod in India*. St Louis: Eden Publishing House, 1930.

Menzel, Emil W. *I Will Build My Church: The Story of Our India Mission and How It Became a Church*. Philadelphia: Board of Missions, Evangelical and Reformed Church, 1943.

Miller, George E. "The Serpent That Failed", *World Call*, July 1922: 28–30.

Nelson, A.E. *Central Provinces District Gazetteer: Raipur District*. Bombay: British India Press, 1909.

"Official Census of India (Reprinted from the Missionary Herald)", *World Call*, July 1924: 37.

Patros, Simon. *Simon Patros of India: A Miracle of God's Grace*. St Louis: Board of International Missions, Evangelical and Reformed Church, Inc., 1954.

Paul, M.M. *Satyanami Panth: Shri Gosain Ghasidas Girod Vasi*. Allahabad: The Mission Press, 1937.

Pickett, J.W. *Church Growth and Group Conversion*. Lucknow: Lucknow Publishing House, 1956.

Richter, Julius. *History of Missions in India*. Translated by Sydney H. Moore. New York: Revell, 1906.

Russell, Robert Vane. *The Tribes and Castes of the Central Provinces of India*. Volume 1. London: Palgrave Macmillan, 1916.

Saum, H.C., and Saum, Mildred. "Followers of the True Name". *Mungeli News Letter*, March 1935: 3.

Sebastian, Mrinalini. "Reading Archives from a Postcolonial Perspective". *Journal of Feminist Studies in Religion* 19, no. 1 (2003): 5–26.

Seybold, Theodore C. *God's Guiding Hand: A History of the Central Indian Mission 1868–1967*. Philadelphia: United Church Board for World Ministries of the United Church of Christ, 1971.

Sharpe, Eric. *Not to Destroy but to Fulfil: The Contribution of J.N. Farquhar to Protestant Missionary Thought in India Before 1914*. Uppsala, Sweden: Almqvist & Wiksells Boktryckeri AB, 1965.

Snow, David A., and Richard Machalek. "The Sociology of Conversion", *Annual Review of Sociology* 10 (1983): 167–90.

Tanner, Th. *Im Lande der Hindus Oder, Kulturschilderungen Aus Indien*. St Louis: n.p., 1894.

Young, Richard Fox. "Some Hindu Perspectives on Christian Missionaries in the Indic World of the Mid Nineteenth Century". In *Christians, Cultural Interactions, and India's Religious Traditions*, edited by Judith M. Brown and Robert Eric Frykenberg, 37–60. Grand Rapids, MI: William B. Eerdmans, 2002.

4

VOICES, TEXTS, AND CONTEXTS IN FILIPINO CHRISTIANITY

Jose Mario C. Francisco

Studies on Christianity in the Philippines have noted its extensive social presence and overwhelmingly high membership at about 80 per cent Catholic of the approximately ninety million total population today (Moreno 2006). This is not surprising, given the four centuries under the sixteenth-century juridical agreement between the Spanish monarchy and the Papacy usually known as the *Patronato Real de las Indias* (Schumacher 1975: 9–11). Christianity's historical development in the Philippine archipelago is thus described in terms of its generally collaborative but occasionally uneasy relations with colonial authorities as well as in terms of the ebb and tide of native resistance to it, especially in the Muslim communities of Mindanao.

A sociolinguistic perspective on religion in the Philippines

The Catholic Church played a prominent role in the initial establishment and subsequent development of colonial society. This was largely through the formation of settlements patterned after European medieval towns (Javellana 2003), and with Catholic institutions involved in health care, social development, and education. With the increasing abuses endemic to colonial power, however, and the rise of nationalist and revolutionary movements in the nineteenth century within and outside the church, leading to the 1898 birth of the Filipino nation, Christianity eventually became institutionally dissociated from Spanish colonialism.

Though these studies mark the broad contours of Christianity's historical development in the colonial Philippines, scholars have paid little attention to the linguistic aspects of the interaction of the native with the Spanish and the Christian, or to the central role that different languages and their usages played in the construction of Filipino Christianity. The few studies that did, notably Ileto's *Pasyon and Revolution* (1979) and Rafael's *Contracting Colonialism* (1988), often concentrated on the semantic; that is, thematic content of religious texts in native languages. This is not surprising, since beliefs about the nature of reality and the significance of life are considered constitutive of religious traditions, especially of the so-called Abrahamic "religions of the book" – that is, Judaism, Christianity, and Islam – and even to a lesser extent, of the Eastern traditions of Hinduism and Buddhism. As a result of this valorization of the semantic and inevitably the textual, sacred writings – whether scripture or sutra – are read and analysed for what they express regarding the teaching of their founder or the doctrine of their disciples.

However, the function and significance of language in religious traditions transcend that of being mere vessels of religious content. According to Ricoeur, these religious traditions, as well as the experiences they mediate, can be "identified through [their] language, or, to speak more accurately, as a kind of discourse" (1995: 35). Like all other social practices, they employ language in multifaceted ways. Religious discourse involves different interlocutors depending on context, and it is often an essential part of many rituals. Sacred texts, from their composition onwards, are themselves inextricably linked with social contexts. Thus a more integral framework for the study of language and religion needs to focus on the interplay of voices, texts, and contexts.

Such a focus one finds in Bauman and Briggs's concept of performance "seen as a specially marked, artful way of speaking that sets up or represents a special interpretative frame within which the act of speaking is to be understood" (1990: 73). Any text is indeed performed in particular contexts through various ways of reading: "performance rather provides a frame that invites critical reflection on communicative processes" (60), and "performance-based research can yield insights into diverse facets of language and their interrelations" (61). In "stress[ing] the cultural organization of communicative processes" (61) but without falling into what Hanks calls the opposite extremes of formalism or sociologism (1989: 100), this frame takes into account various dimensions of social context, such as the literary and the institutional, and the role they play in the performance of a text.

Within this general framework, the present essay revisits the development of Filipino Christianity and focuses on the complex interplay of voices, texts, and contexts. In particular, it highlights the dynamic between texts and their oral performance in particular social contexts and suggests that this dynamic provides the sociolinguistic basis for Christianity's enduring and multifaceted influence in Philippine society. Its body consists of two major parts: The first section discusses the transformation of native languages through the romanization of native scripts and the textualization of native religious discourse, while the second section focuses on the entextualization of native religious texts, illustrated by the performance and the text of the Philippine narrative of Christ's Passion.

"And the word was made text"

The introduction of late-medieval post-Reformation Spanish Catholicism shaped native society through its transformation of native languages and modes of discourse. In order to pass on Christian doctrine and practice to natives tempted to return to their "pagan" ways, the colonial church – influenced by the more benevolent Spanish views towards natives of Juan de Vittoria and others – decided, at the Synod of Manila, convened in 1582, to use local languages for preaching (Schumacher 1975: 75). Such a missionary strategy established a complex economy between and amongst the colonial church's languages (Greek, Latin, and Spanish) and the local languages. Furthermore, it led to the creation of native religious discourse for the sake of the colonial church's self-understanding, as well as for the communication of Christianity to the *indio* (the Spanish term for the indigenous peoples).

Romanization of the native script

These local languages, most probably descended from the Austronesian family (Scott 1994: 12–4), had to be altered first. Their scripts consisted, in Goody's estimation, of a form of South-Indian syllabic script similar to those of other south-east Asian societies (1993: 51) and, according to some contemporary missionary accounts, occasionally varied in form from one island community to another (Scott 1994: 95). The Tagalog *babayin* "contained seventeen letters, three of which were the vowels *a*, *e-i*, and *o-u*" and "the other fourteen were radical signs representing the consonants" (Scott 1994: 213–4). Because the pronunciation of syllables depended on putting diacritical marks (*kudlit*, in Tagalog) on the characters for consonants, syllables could not end in consonants and could be read in different ways, the correct one known only through the context in which it was used. Furthermore, the direction of inscribing, and thus reading, such syllables could be either horizontal or vertical (Scott 1994: 215).

An existing script similar to these pre-Hispanic languages is preserved in the *ambahan* poetry of the Hanunoo-Mangyans, an indigenous people of Mindoro island. These short lyrical poems are "scratched into the surface of a piece of bamboo, with a sharp pointed, home-made knife, and with this memory aid it was so much easier to remember and learn by heart the traditional poetry" (Postma 2005: 7–8).

Old texts in this native script are attested to by historical sources like the Boxer Codex, but they were short pieces intended as messages, pronouncements, or contracts (Scott 1994: 209–16). The Laguna Copperplate Inscription – the earliest (tenth century) extant text using Sanskrit, Old Javanese, Old Malay, and Old Tagalog – was an official proclamation by the chief of Tundun that he was condoning a debt worth about a kilogram of gold incurred by someone named Namwaran (Tiongson 2008).

The brevity of these texts are linked to the very nature of native scripts. As Scott declares, "the inability of the baybayin to represent the *ng* of *nitong* illustrates its inadequacy for recording spoken Tagalog faithfully, probably a sufficient explanation for its limited use and the absence of Filipino documents, written records, or monumental inscriptions" (Scott 1994: 215). It thus becomes clear that native society, whether pre-Hispanic or contemporary like the Mangyan, is rightly characterized as a traditional oral-aural culture, wherein what Ricoeur calls "the dialectic of event and meaning" in discourse remains tied to its particular interlocutors and context (1978: 8–12; 29). Its languages served as the medium for social intercourse within and amongst small local settlements (*barangays*) characterized by fluid social relations (Scott 1994: 12–4, 95–6, 186, 209–16). Their linguistic practices correspond to those observed by Ong in such cultures: proverbs and riddles transmitted commonplaces ("loci communes"), songs accompanied many ordinary activities such as boating, and narratives such as epics were told in communal feasts (1981: 22–35). All these indicate not only a specific set of behaviour but a particular way of being: "For oral-aural man [*sic*], utterance remains always of a piece with his life situation. It is never remote" (33).

But those who were "outsiders" in native society, like the Spanish missionaries, found reading even such short texts difficult. Augustinian Gaspar de San Agustin, who later wrote the *Compendio de la lengua tagala*, complained that "the script is as easy to write as it is difficult to read because it is to guess [*adivinar*] it" (1787: 155). If local languages were to be the medium for preaching, the native syllabary would have to be transposed and codified into the roman alphabet, "an organized system of writing, not an assortment of more or less isolated signs" (Ong 1981: 36). With this phonetic transposition of these languages, "the alphabet irrupts into sound itself, where the one-directional flow of time asserts its full power" (Ong 1967: 45).

This seemingly straightforward transposition of native scripts into the roman alphabet literally opened a new world and had far-reaching consequences. Local languages were not only made accessible to any non-native speaker, but also codified in grammars and standardized with dictionaries. Furthermore, native discourse could now be committed to writing and propagated through the technology of printing, initially through xylographic means and in the seventeenth century with movable printing presses (Javellana 2010).

Thus countless manuscripts and publications were produced for missionaries and other functionaries of the colonial establishment. Examples are grammars such as Agustin de Magdalena's *Arte de la Lengua Tagala Sacado de Diversos Artes* (1679), and painstakingly compiled dictionaries such as the best-known *Vocabulario de la Lengua Tagala* (1754) completed by Dominicans and Jesuits (Lumbera 1986: 30–46). Sermon anthologies by Dominican Francisco Blancas de San Jose, known as the local Demosthenes for his mastery of Tagalog and Chinese (Francisco 1994), catechisms like Franciscan Juan de Oliver's *Doctrina Cristiana* (Cruz 1994), and even confession manuals like Sebastian Totanes's *Manual Tagalog* (Rafael) proved useful materials for newly arrived missionaries not yet fluent in the local languages.

Textualization of native religious discourse

But of far greater significance were texts written for native speakers. Iconic amongst them is the first book printed, Franciscan Juan de Plasencia's 1593 *Doctrina Cristiana* (1991). It contains the Tagalog alphabet in roman script as well as basic Christian prayers and doctrines in Spanish, romanized Tagalog, and the old syllabic script. Missionary Francisco Lopez explains why he incorporated native script in his doctrina:

> the reason for printing the text of the Doctrina in Tagalog type . . . has been to begin the correction of the said Tagalog script, which as it is, is so defective and confused (because of not having any method until now for expressing final consonants – I mean those without vowels) that the most learned reader has to stop and ponder over many words to decide on the pronunciation which the writer intended.
>
> (1895 [1621]: 92)

Thereafter followed pamphlets (*caton*) through which natives learned basic Christian prayers, as well as the roman alphabet of their local languages.

The textualization of native discourse came through this integration of religion and literacy. Its consequences were far-reaching, as Goody pointed out in his study of the impact of literacy on various religions: "Literate religions have some kind of autonomous boundaries. Practitioners are committed to one alone and may be defined by their attachment to a Holy Book, their recognition of a Credo, as well as by their practice of certain rituals, prayers, modes of propitiation" (1986: 4–5). As a result, "literate religions . . . at least alphabetically literate ones, are generally religions of conversion, not simply religions of birth" (5).

Thus these religious books became the first truly native texts to constitute, in Hanks's terms, "a configuration of signs that is coherently interpretable by some community of users" (Hanks 1989: 95). Given the status of these native texts, both as code and work "centered. . . [or] grounded in a locally defined social context" (106), natives learned not only how to read their own languages in roman script, but also how to practise their new religion. In the words of the Augustinian Agustin de Magdalena, "this *Arte* is an instrument for the use of the Ministers of the Faith who desire to preach to the pagans, that they may have persuasive and clear voices [*vozes vivas y claras*] with which to manifest God's doctrine" (1679: unpaginated).

These native texts emerged from and were read in context or, as Hanks put it, "voiced" in context (1989: 102). Hence, attention must be shifted "from the text considered in isolation to the active engagement of a reader in constituting the text itself, thus raising interesting questions about reader response, reception and the balance between what is said in the text, what is unsaid but available, and what must be constructed" (97).

These elements are evident in how these and subsequent texts were produced and performed. Like the *catons* intended to foster devotion and instruction of the natives, prayers and novenas with brief stories about saints sourced from Europe and the New World were translated. Their performance was often linked to communal religious activities, such as catechetical classes for the youth, praying over the dying in far-flung villages, and parish devotions. Some took place inside the church under the direct supervision of the local priest; others took place outside, in the homes of local Christians or even in the streets, as in the case of processions. What the missionary Francisco de San Antonio describes is typical:

> Every Saturday in our missions the Mass of the Immaculate Conception is sung with all possible solemnity, and the *Salve* [Hail, Holy Queen] in the afternoon before our Lady's altar. After the Mass, the *baguntao* and *dalagas*, that is, the young people of both sexes, recite the catechism out loud under the direction of their *fiscal* or supervisor – an office created by royal ordinance and invariably occupied by a native chief selected from among those in highest esteem.
>
> (Schumacher 1979: 48)

As the account indicates, these practices required the services of literate native lay leaders (*fiscales*). Despite the absence of comprehensive studies or materials on the extent and level of literacy amongst the natives, missionary accounts attest to their desire for literacy:

> [The native's] love for books is so great that, not satisfied with those printed in their language, written by religious men, with sermons they hear, and with bible histories, lives of saints, prayers, and sacred poetry composed by themselves, there is hardly any – much less a woman – who does not have one or more books in their language and script written by themselves, a thing unknown among neophytes in any other nations.
>
> (Santiago 2003: 561)

Though such accounts might have been exaggerated, some local functionaries of church or government were highly literate. One of these was Juana de Sanct Anttonio, the first known native woman writer, who wrote theological reflections in the mid-1660s (564–5).

Given such regard for literacy, religious texts were highly valued and became "repositories of power" (Cannell 2006: 158–9). *Fiscales*, often entrusted with the physical possession of the religious books, "became both the literal and metaphorical brokers of Christian meaning" (137). Even Latin and Spanish words from these texts served as sacred objects for the 1890s Colorum sects, which took their name from the Latin prayer ending "*per omnia saecula secolorum*" ('forever world without end') (Ileto 1979: 93) or in the *anting-anting* ('amulets') inscribed with corrupted Latin and sold outside the popular Quiapo church today.

Thus the romanization of the native script and the subsequent textualization of native religious discourse turned into, borrowing Ricoeur's words, "a tremendous cultural achievement" (1976: 26) and made discourse "more spiritual in the sense that it is liberated from the narrowness of the face-to-face situation" (31). With this indication of "the textuality of Christian Faith" (Shorter 1992: 3) and Christianity as "a vernacular translation movement" (Sanneh 1989: 7), Filipino Christianity emerged as a result of the encounter between Spanish Catholicism and native culture and was compared by Rafael (1988: 211) to 'translation' (Sp: *traducir*):

> conversion in early Tagalog colonial society was predicated on translation; yet the Spaniards' and the Tagalogs' notions and practices of translation differed to the degree that the relative position of one to the other remained ambiguously defined. Christian conversion and colonial rule emerged through what appeared to be a series of mistranslations . . . such mistranslations were ways to render the other understandable.

Transcending intentions and expectations on both sides, this "carrying over" did not involve finding literal or functional equivalences for words, as some understand translation to be; instead, it involved mediating between social worlds expressed in language through "an ongoing process of cultural transfer" (Bassnett and Trivadi 1999: 2).

For instance, these native religious texts used common Tagalog words in a Christian context, incorporated Spanish Christian words into native discourse, and employed native religious terminology for Christian realities. Catechisms drew from local experience, and traditional literary forms such as folk songs were used for prayers to saints. Stories of saints were told in traditional verse forms.

This dynamic mediation between social worlds expressed in language engendered Christian doctrines codified in native vocabulary and laden with local connotations. For instance, Franciscan Blancas' sermons on the classical doctrine of redemption from the slavery of sin took on different nuances, because the native practice of slavery differed from chattel slavery in feudal Europe (Francisco 2007: 70–5). In pre-Hispanic native society, "an *alipin* [slave] was a man in debt to another man. His subordination was therefore, obligatory, not contractual: the other man was technically his creditor rather than his lord" (Scott 1983: 146). Moreover, "[these] forms of indebtedness and servitude . . . were transferable and negotiable, allowing for random submission to authority . . . rendering tribute [tax] and performing labor were less signs that memorialized one's submission to the master than ways of bargaining with him" (Rafael 1988: 145).

This native view and practice of slavery, based not on a formal legal code (as the colonizers might have expected) but on a more flexible system of transaction related to indebtedness, molded the native reception of Christian redemption. Negotiation became characteristic of the relations between God and humanity, described in a sermon as *maquipagsaysay*, literally, 'settling one's accounts' (Francisco 1994: 515). Because of such an image of a master willing to negotiate, God was more easily seen as "Lord God who turns to, runs after and forgives his sinner-slaves who offended him terribly" (89) and who "woo[es] his slaves" (369). Within this view of a close divine-human relationship characterized by negotiation, God's magnanimity became even more pronounced. As master, God was under no obligation to save sinful humanity by sending his son. As Jesus Christ the Saviour, God took up the life of a slave freely. This double magnanimity – when read within the Tagalog social context – signifies a patronage not born out of legal obligation or social contract, as in feudal society, but based on sheer favour. But beyond particular beliefs, given native nuances, Filipino Christianity offered a paradigm for social and personal redemption based on imitating Christ and the saints, who were characterized with native sensibilities in native religious texts. Thus, in bringing the Christ story into native hearts and minds, these texts transformed local religious discourse into a language of redemption and created a Christianity with native meanings, appropriated by natives through popular practice.

The contours of Filipino Christianity have been discussed in Ileto's study of nineteenth- and early-twentieth-century social movements. Empowered to see events in the light (*liwanag*) of the Christ story, these movements acted with great resolve (*matibay ang loob*) in solidarity (*damay*) with Jesus's life unto death (1979: 65–72). Their rituals often prefigured a heavenly order free from social inequalities. For example, Hermano Pule, a native founder of an early–nineteenth-century commune, described the attributes of those in heaven using Thomistic ideas and envisioned their relations as equality, the exact opposite of earthly differences in appearance, intelligence, wealth, and status (46–52).

"And the text was made word"

Filipino Christianity emerged and took root in the local landscape not only through the production of native religious texts, but also because of the diverse ways and contexts in which these texts were read. Bauman and Briggs (1990) explained this in terms of entextualization – "the process of rendering discourse extractable, of making a stretch of linguistic production into a unit – a *text* – that can be lifted out of its interactional setting" (73). Thus "decontextualization [of a text] from one social context involves recontextualization in another" (74), and these two are but "two aspects of the same process, though time and other factors may mediate between the two phases" (75).

The significance of this process becomes more evident when framed in relation to performance. Reading is, after all, performing a text. Given this, Hanks (1989) rightly pointed to "the phenomenon of *voicing in text*" which "subsumes the relation between the textual artifact and the framework of production and participation from which it arises" (102). This performative dynamic points to "the sociocultural encounter between text and audience (and therefore author and audience), the institutional and ideological formation of ways of reading, the imposition of interpretation, the lack of closure in text, the dialogical construction of meaning, and the functions of reader expectations in centering" (112). This focus on the essential link of entextualization or reading of a text in a different context with the reader's performative voicing of and in the text underscores the point that "the decontextualization and recontextualization of performed discourse bear upon the political economy of the text" (Bauman and Briggs 1990: 76).

Entextualization of native religious discourse

Economic and cultural changes during the Spanish colonial period created the need for new texts for and by native voices. Increased trade with the 1834 opening of Manila's port stimulated domestic agricultural production and encouraged land ownership and use (Legarda 1999: Ch. 4). This generated wealth and power amongst those already prominent, such as municipal officeholders (*principalia*), and also amongst Spanish or Chinese mestizo and native families who desired to equal Spaniards in urbanity and learning (Cullinane 2003: Ch. 1).These changes in social context brought about corresponding changes in ethos, such as the need for literature other than directly intended for worship or catechetical instruction. Though Spanish church and colonial authorities owned printing presses and regulated the publication and dissemination of manuscripts (Javellana 2010), literate natives, some of whom even knew some Spanish, produced narratives or books on Christian etiquette.

Most popular amongst this new literature were metrical romances (*auit*) imported from Mexico without church interference and adapted by laypersons from Spanish medieval stories about Christians and Moors. These stories typically charted convoluted loves and political intrigues in kingdoms imagined by their native authors.

One such *auit* is native Francisco Balthazar Balagtas's well-known *Florante at Laura* (1861), in which Florante is helped by a Moor to save Laura and to restore order 'within and around my wretched land [where] betrayal reigns supreme (*Sa loob at labas ng bayang cong saui/kalilua'y siyang naghahari*)' (Ronquillo 1921: 8, stanza 14). The kingdom's struggle for justice and Florante's desire for Laura reflect the Christian passover from suffering to liberation, expressed in terms of a cosmic order governing social and personal affairs. Characters invoke 'merciful heaven (*maawaing Langit*)' and act according to 'reason (*catouiran*)' or 'natural law (*natural na ley*)' in their quest for love and justice (Francisco 1989: 129–40). Both in its description of tyranny and in its vision of well-being, Balagtas's *auit*, when read within its historical context, was a critique of the colonial situation and an expression of the desire for change.

Changes in the contexts out of which these new texts emerged corresponded to changes in the contexts in which they were read. While earlier texts were read within the confines of the church complex, most of these new texts were read in locations outside direct ecclesiastical control, such as private houses or along the streets. For instance, Rafael points out how the performance in the town plaza of dramatized metrical romances (*comedia*) contributed to the creation of public space in the colonial landscape (2006: Ch. 6). These performances, together with other religious practices which took place in public places, provided occasions for natives to gather. This eventually led to the increasing scope of civil space for nationalist and revolutionary movements in the late nineteenth century.

Though not explicitly devotional like *novenas*, these newer texts and the various contexts in which they were read were based and embellished the Christ story. They might have been about how rulers struggled for power in faraway kingdoms or how young men and women ought to behave according to Christian manners, but they sought nonetheless to re-appropriate the Christ story in different situations. These re-appropriations of the Christ story took many forms. Sometimes, they were simple applications of the Christ story to new areas, such as parent-child relations in more urbane settings; other times, they took on nativist or millenarian characteristics like Felipe Salvador's religiopolitical movement called *Santa Iglesia* (Holy Church) which flourished from 1894 to 1910 (Ileto 1979: 259–313). But whatever their form, these religious texts based on the Christ story captured the native imagination – in part, because in them Christianity took on a native form through the transformation of native discourse and, in part, because local priests were insufficient in number to accompany the native population in their faith journey.

Performative reading of the pasyon text

The entextualization of native Christian discourse is best illustrated by the native tradition of the Christ story which begun during the early colonial period but continues to be vibrant in contemporary Philippine society. This tradition's magisterial text remains the 1812 *Caysaysayan ng Pasiong Mahal* (*Story of the Holy Passion*), also known as *Pasyong Pilapil* (*Pilapil's Passion*) or *Pasyong Henesis* (the *Genesis Passion*) (Javellana 1988). Its textual history reflects its entextualization in changing contexts. Native Aquino de Belen's 1704 text, originally published with prayers for those about to die, constitutes its narrative core from Jesus's Last Supper to his Resurrection (Javellana 1990). Under the supervision of Fr Mariano Pilapil, the *Casaysayan* updated the vocabulary and theology of the earlier text and framed this updated text between Creation and the Last Judgement to express the sweep of salvation history.

These textual differences between Aquino de Belen's and the *Casaysayan* are reflected in the social contexts in which they were performed – the earlier read by native lay leaders praying over the dying at their homes, and the *Casaysayan* chanted during Lent, the liturgical season commemorating Christ's Passion, Death, and Resurrection, especially Holy Week. Revisions of the *Casaysayan* text followed for various reasons during the late nineteenth and early twentieth centuries. In some regions, original manuscript versions of the *pasyon* were commissioned by prominent families and regarded as heirloom pieces. But the popularity of the *Casaysayan* prevailed, making it the local book published in greatest number of copies, editions, and translations into the major languages of the Philippines. This popularity was in no large measure due to the manner in which this text was performed. While the *pasyon* text could be read alone as devotional prayer, its natural context was social. The text came alive through communal practices of antiphonal chanting called *pabasa* ('reading').

This performative reading of the *pasyon*, often done as a family tradition or in fulfilment of devotional promise (*panata*) for a particular blessing, such as recovery from illness, has undergone little change through the centuries. Though slight variations occur from region to region or on the basis of the social status of those involved, the *pasyon* text is usually chanted before an altar in a private house during Lent. This occurs on days with some Christian significance like Friday and Sunday, but during Holy Week, the chanting can go on all day and all night amidst the usual activities of the household. Moreover, the chanting is broadcast over the neighbourhood with loudspeakers when the organizers have the means. Thus distinctive of the *pabasa* is its overwhelming presence, no matter what one is engaged in, one hears its performance literally enveloping the entire household and even neighbourhood. This presence is intended so that all might hear the story of Christ's passion and thus learn from it. Noteworthy too is its mode of chanting. Family members, neighbours, and others from afar gather and take turns in the antiphonal chanting of the *pasyon* text, alternating stanzas. Each group of chanters draws from and improvises with traditional melodies or, when the young participate, contemporary pop music. This interplay is an essential element in the *pabasa*, as Cannell (2006: 143) discovered in her fieldwork in the Bicol area of southern Luzon:

> they [chanters] themselves emphasize their skill in producing the words clearly. . . . Yet at the same time, they stress the 'matching,' 'harmony,' and 'pairing' of their two voices, concepts that refer to ideas of balance, testing, and blending in Bicol rhetorical techniques in quite different contexts (such as formal riddling contests and courtship ritual).

Cannell found haunting "the 'pairs' of readers catching each other's eye to coordinate a complex point in the ornamentation, expressions transformed by the act of shared concentration, by the use of a skill and a genre so well mastered as to allow the pleasure of improvisation" (252).

Thus the performative reading of the *pasyon* in the context of family residences erased boundaries between participant and spectator as well as between the religious and the social. As Bauman and Briggs explained, "contextualization involves an active process of negotiation in which participants reflexively examine the discourse as it is emerging, embedding assessments of its structure and significance in the speech itself" (1990: 69).

Orality in the pasyon *text*

The orality of the *pasyon* tradition is reflected not only in its performative reading in changing contexts, but also in the oral character of the imbedded roles of narrator and listener in the *pasyon* text itself. The dynamic between these roles supports the oral nature of the tradition and therefore magnifies the impact of its performance. Unlike the often hidden narrator in biblical texts, the *pasyon* narrator is overt, recounting events in the Christ story and addressing the listener directly in certain sections of the text. From the *pasyon*'s opening prayer to God onwards (Javellana 1988), the narrator seeks to be identified with the listener, both being Christian and native. The personal pronoun which the narrator uses as an address is the Tagalog inclusive first-person plural, *tayo* (128, stanza 2652). The listener is called by the narrator *capatid co* ('my sibling'), with the listener representing other baptized natives (71, stanza 853).

However, the narrator and the listener differ in one significant aspect. On one hand, the narrator is consistently portrayed as a converted native, tested in the faith, deferential to Scripture or church teaching, and, being omniscient, credible as witness to events in Jesus's life. Thus is the narrator able to provide commentary through the moral lesson following important events in Jesus's life or the choice of events and the way these events are described. On the other hand, the listener or reader is presented as "a enslaved Christian (*cristianong nagugupiling*)", one enslaved to sinfulness and thus in dire need of conversion (76, stanza 1004). Thus the relationship between the narrator and the listener appears to be nothing more than an instance of one native Christian preaching to another. This relationship between narrator and listener, however, is transformed because of the dynamic in its antiphonal mode of performance. Because stanzas are chanted alternately, no group – and therefore, no individual – assumes the voice of the narrator or the listener exclusively. The overall effect, then, is that each and every one involved becomes both narrator and listener. Chanters, individually and as a group, are both narrators and listeners and, therefore, preacher and preached-to.

This shifting of roles becomes an effective rhetorical device to promote repentance and conversion among all involved in the *pasyon*'s performance. What chanters experience is a continuous shift in roles: in one stanza, they chant as narrator calling sinners to repentance; in the next, they listen to the other group of chanters.

This device is also at work in the lengthy passages of dialogue in the *pasyon* text. When these passages are chanted, the chanter assumes the roles of different characters: a wailing Mary, a sly Herod, a despairing Judas, and even a pleading Jesus. This shifting of roles may be analogous to what Iser (1978: 156–7) described as the split within the reader and its subsequent impact:

> The split between subject and himself, which results in a contrapunctually structured personality in reading, not only enables the subject to make himself present to the text, it also brings about a tension, which indicates to what extent the subject has been affected by the text.

By having the chanter assume the roles of narrator and narratee, as well as those of the other characters, the *pasyon* evokes from the chanter participation at many levels. The chanter literally speaks and appropriates the words of various characters – both good and evil – as well as those of the narrator and the listener. All these words are chanted by the same voice of the chanter, and thus preaching is completely internalized. No longer is the preacher an outsider nor the penitent a passive listener to the dramatic story of salvation played in the present within the chanter. Both are now one in the chanter, who preaches to himself or herself.

Religious voices in the town plaza

The orality of the *pasyon* tradition embedded in its magisterial text, as well as expressed in its performance in many contexts, illustrates the significance of the oral in the nature and development of Filipino Christianity. Spanish Catholicism brought literacy to a primarily oral native society and transformed its culture through the romanization of its native script and textualization of native religious discourse. These religious texts were written by natives in their now romanized languages, to be read in social settings involving family and community. Moreover, because of their nature as texts, they could and were read in changing contexts through the process of entextualization.

This dynamic between voices, texts, and contexts provided the linguistic foundation for the emergence of Filipino Christianity. Christianity could not have taken this native form if local languages had not been used in religious discourse. When employed to express Christian faith, these languages brought to expression a world of meanings involving images, feelings, and concepts rooted in the native society. Furthermore, with the performance of these native religious texts in various contexts, this world of meanings was given multiple voices.

Revisiting the development of Filipino Christianity through a sociolinguistic perspective has created a pathway to current discussions between orality and literacy. Spunta observes that "from the time of Antiquity, preference has been given in turn to orality and literacy" (2006: 93). On one hand, "orality has been repeatedly praised as the 'original', pristine form of communication, whose fragile 'image' is to be found in writing"; on the other, "the so-called 'alphabetic bias' of modern Western societies . . . values literacy at the expense of orality, which it views as a more primitive form of communication, reflecting a less advanced state of social and cultural development". (93)

However, more recent research "has moved towards a more flexible, encompassing cross-modal model of communication, which recognizes the relative – rather than fixed – relationship between speech and writing" (Spunta 2006: 93). Caesar and Spunta point to "the revival of the interest in the oral . . . as an indirect response to the increasing oralization of contemporary culture, and the growing perception of that trend, which Ong dubbed 'secondary orality in America in the 1950s and 1960s, and in Italy later, from the late 1970s on" (2006: 1).

The historical development of Filipino Christianity exemplifies the fluid boundary between the oral and the written. Romanization of native languages and textualization of native religious discourse produced and circulated religious texts which were intended to promote conversion and safeguard the purity of Christian faith through censorship by church authorities and control of the means of publication. However, because these texts were performed in familial and social spaces that were beyond official church and colonial authorities, reading these texts gave rise to meanings – whether critical or not – unintended and unforeseen by these authorities. The particular example of the *pasyon* tradition provides an exemplar of this interplay between the oral and the written. Since its origin in Aquino de Belen's 1704 poem, its text was constantly revised until the magisterial 1812 *Casaysayan*. However, the modes of performance of this text, as well as the role-shifting between narrator and listener embedded in the text, exemplify, on one hand, respect for the text and, on the other, variation in its reading. All these create an ethos in which preaching is internalized within the native Christian subject, and conversion, that defining characteristic of literate religion (Goody 1986: 5), is promoted. Moreover, because the performative reading of this text occurs in contexts diverse in space as well as time, the Christ story which it offers as paradigm for personal and communal redemption is heard *within* these contexts. Hearing the Christ story within these diverse and changing contexts is what made the emergence and development of Filipino Christianity possible.

The oral and the textual in Filipino Christianity, far from competing with each other, then, serve each other for the sake of preaching the Christ story in context. Moreover, the preaching of this story through performative reading, illustrated by the *pasyon* tradition, gathers community. As Ong (2001: 74) explains:

> because of its physical constitution as sound, the spoken word proceeds from the human interior and manifests human beings to one another as conscious interiors, as persons, the spoken word forms human beings into close-knit groups. When a speaker is addressing an audience, the members normally become a unity, with themselves and with the speaker.

Though this is especially true for Christianity, which proclaims that "faith comes through hearing" (Letter to the Romans 10, 17), the *pasyon* tradition's power to gather people through chanting is magnified on two counts: first, the interpersonal link between narrator/preacher and listener/preached-to among the alternating chanters and, second, its enveloping presence over household and neighbourhood while being chanted. This dynamic between orality and literacy remains characteristic of Filipino Christianity today. Though forces of modernization, urbanization, and globalization have effected change in Philippine society, Christian practices such as processions, patronal feasts, and popular rituals continue, albeit modified by these forces. Many urban poor communities hold the Lenten practice of the Stations of the Cross, during which they link Jesus's path to Calvary with their social concerns. Less than a stone's throw from Manila's international business district lies Makati *viejo* ('Old Makati'), site of vibrant Holy Week practices including the *pabasa*, the Good Friday procession of crosses and the Easter midnight's reenactment of the Virgin Mary's encounter with the Risen Jesus. Even the widespread presence of electronic media is employed in the service of the dynamic between orality and literacy. El Shaddai, the most popular Catholic charismatic group, led by Mike Velarde, integrates radio and television with mass rallies and home meetings and, through this integration, has created a new form of community. According to Wiegel (2005: 48–9), the radio and TV broadcasts extend the ritual sphere of the *Gawain* (home meetings) beyond its immediate locale because, before and after the event, radio and TV are played constantly. Since the broadcast is live, one begins experiencing the event even while still at home. Within the PICC grounds, radios serve as links with Brother Mike at the centre, focusing attention on Velarde and the events onstage.

The mass-mediated community is gradually transformed into the immediate, physical community of the rally.

After the rally, the opposite process takes place: the dispersing crowds remain glued to continuing radio transmissions of El Shaddai programmes, and "the mass rally congregation, even today, is in effect an extension or continuation of the radio listening audience, and the borders between the two types of experience are blurred rather than distinct" (Wiegel 2005: 508). This dynamic interplay results in a blurring of boundaries between mass-mediated communications and gatherings in large public or small local venues, creating an atmosphere participants refer to as "live". This 'live' atmosphere is in no small measure cultivated by Velarde's style of talking – "conversational, direct, never losing track of the person who is listening" – so much so that it "contributes to an image of community that is close and familiar" (56). Moreover, local gatherings, formal or informal, often use replays of El Shaddai radio and television programmes, and voice recordings of Brother Mike circulate among Filipino migrants and overseas workers (53). It is not entirely surprising that Velarde's voice is frequently heard on many issues confronting contemporary civil society. But his is not the only one – more historic is Cardinal Jaime Sin's voice on radio calling all Filipinos to the 1986 People Power Revolution against the Marcos dictatorship. Velarde's, Cardinal Sin's, and many other Christian voices would not be heard today without the vibrant dynamic between orality and literacy in Filipino Christianity.

References

Baltazar, Francisco Balagtas. *Pinagdaanang buhay ni Florante at Laura cahariang Albania* [Life experienced by Florante and Laura in the kingdom of Albania]. Edited by Carlos Ronquillo. Manila: Imprenta de Ramirez y Giraudier, 1921 [1861].

Bauman, Richard, and Briggs, Charles L. "Poetics and Performance as Critical Perspectives on Language and Social Life", *Annual Review of Anthropology* 19 (1990): 59–88.

Cannell, Fanella. *Power and Intimacy in the Christian Philippines.* Quezon City: Ateneo de Manila University Press, 1999.

Cannell, Fanella. "Reading as Gift and Writing as Theft". In *The Anthropology of Christianity*, edited by Fanella Cannell, 158–59. Durham and London: Duke University Press, 2006.

Cruz, Jose M., editor. *Declaración de la Doctrina Cristiana en Idioma Tagalog* [Declaration of Christian doctrine in the Tagalog language]: *Juan de Oliver OFM (+1599)*. Quezon City: Pulong, Sources for Philippine Studies, 1994.

Cullinane, Michael. *Ilustrado Politics: Filipino Elite Response American Rule, 1898–1908.* Quezon City: Ateneo de Manila University Press, 2003.

Francisco, Jose Mario C. "*Di tuwid ang katwiran: ang etiko't panrelihiyong pananaw sa Florante at Laura* [Reason is not straightforward: ethical and religious views in Florante and Laura]". In *200 Taon ni Balagtas* [200 years of balagtas], edited by Soledad S. Reyes, 129–40. Quezon City: Balagtas Bicentennial Commision, 1989.

Francisco, Jose Mario C. "The Christ Story as the Subversive Memory of Tradition: Tagalog Texts and Politics around the Turn of the Century". In *Text/Politics in Island Southeast Asia: Essays in Interpretation*, edited by D.M. Roskies, 82–110 (Ohio University Monographs in International Studies, Southeast Asia Series, No. 91). Athens, OH: Center for International Studies, 1993.

Francisco, Jose Mario C., editor. *Sermones: Francisco Blancas de San Jose O.P. (+1614)* [Sermons]. Quezon City: Pulong, Sources for Philippine Studies, 1994.

Francisco, Jose Mario C. "Translating Christianity into Asian Tongues: Cultural Dynamics and Theological Issues", *Asian Christian Review* 1, no. 2 (2007): 70–84.

Goody, Jack. *The Logic of Writing and the Organization of Society*. Cambridge: Cambridge University Press, 1986.

Goody, Jack. *The Interface between the Written and the Oral*. Cambridge: Cambridge University Press, 1993.

Hanks, W.F. "Texts and Textuality", *Annual Review of Anthropology*, 18: 95–127.

Ileto, Reynaldo Clemeña. *Pasyon and Revolution: Popular Movements in the Philippines, 1840–1910*. Quezon City: Ateneo de Manila University Press, 1979.

Iser, Wolfgang. *The Act of Reading: A Theory of Aesthetic Response*. Baltimore and London: Johns Hopkins University Press, 1978.

Javellana, Rene B., editor. *Casaysayan nang pasiong mahal ni Jesucristong Panginoon natin na sucat ipag-alab nang puso nang sinomang babasa* [Story of the Holy Passion of Jesus Christ Our Lord to enflame the heart of the reader]. Quezon City: Ateneo de Manila University Press, 1988.

Javellana, Rene B., editor. *Mahal na passion ni Jesu Christong Panginoon natin na tola ni Gaspar Aquino d Belen* [Holy Passion of Our Lord Jesus Christ in verse by Gaspar Aquino de Belen]. Quezon City: Ateneo de Manila University Press, 1990.

Javellana, René, S.J. The Tenacious Village. Paper presented at the Colloquium on Nations, Cities, Civilizations: Stories and Arguments, at Quezon City, Ateneo de Manila University, 4 February 2003.

Javellana, René, S.J. Missionary Printing in the Philippines. Paper presented at the Conference on Galleons and Globalization: California Mission Arts and the Pacific Rim at San Francisco, University of San Francisco, 24 September 2010.

Legarda, Benito J. Jr. *After the Galleons: Foreign Trade, Economic Change and Entrepreneurship in the Nineteenth-Century Philippines*. Quezon City: Ateneo de Manila University Press, 1999.

Lopez, Francisco. *Libro a naisurátan ámin ti bagás ti doctrina Christiana nga naisurat iti libro ti Cardenal a agnagan Belarmino* [Book where all the contents of Christian doctrine are told according to the cardinal named Bellarmine]. Manila: Publisher unknown, 1895 [1621].

Lumbera, Bienvenido L. *Tagalog Poetry 1570–1898: Tradition and Influences in its Development.* Quezon City: Ateneo de Manila University Press, 1986.

Magdalena, Agustin de. *Arte de la lengua tagala sacado de diversos artes* [Art of the Tagalog language taken from diverse arts]. Manila: Publisher unknown, 1679.

Moreno, Antonio F. *Church, State, and Civil Society in Postauthoritarian Philippines: Narratives of Engaged Citizenship.* Quezon City: Ateneo de Manila University Press, 2006.

Ong, Walter. *The Presence of the Word: Some Prolegomena for Cultural and Religious History.* Minneapolis: University of Minnesota Press, 1981.

Ong, Walter. 2001. *Orality and Literacy: The Technologizing of the Word.* London and New York: Routledge, 2001.

Plasencia, Juan de. *Doctrina Christiana en lengua espanola y tagala* [Christian doctrine in Spanish and Tagalog]. Manila: National Historical Institute, 1991 [1593].

Postma, Anton. *Mangyan Treasures: The Ambahan: A Poetic Expression of the Mangyan of Southern Mindoro, Philippines*, 3rd revised edition. Mansalay: Mangyan Heritade Center, 2005.

Rafael, Vicente L. *Contracting Colonialism: Translation and Christian Conversion in Tagalog Society under Early Spanish Rule.* Quezon City: Ateneo de Manila University Press, 1988.

Rafael, Vicente L. *The Promise of the Foreign: Nationalism and the Technics of Translation in the Spanish Philippines.* Pasig City: Anvil Publishing, 2006.

Ricouer, Paul. *Interpretation Theory: Discourse and the Surplus of Meaning.* Fort Worth: Texas Christian University Press, 1976.

Ricoeur, Paul. "Philosophy and Religious Language". In *Figuring the Sacred: Religion, Narrative, and Imagination*, edited by Mark I. Wallace, translated by David Pellauer. Minneapolis: Fortress Press, 1995.

Ruiz, Miguel. *Bocabulario tagalo* [Tagalog dictionary], edited by Jose Mario C. Francisco. Quezon City: Pulong: Sources for Philippine Studies, 1997.

San Agustin, Gaspar de. *Compendio del arte de la lengua Tagala* [Compendium of the art of the Taglog language]. 2nd edition. Sampaloc: Convento de la Nuestra Señora de Loreto, 1787 [1703].

Sanneh, Lammin. *Translating the Message: The Missionary Impact on Culture.* New York: Maryknoll, Orbis Books, 1989.

Santiago, Luciano. "The Flowering Pen: Filipino Women Writers and Publishers during the Spanish Period, 1590–1898, a Preliminary Survey", *Philippine Studies* 51, no. 4 (2003): 561.

Schaeffner, Christina, and Adab, Beverly. "Translation as Intercultural Communication – Contact as Conflict". In *Translation as Intercultural Communication*, edited by Mary Snell-Hornby et al. Amsterdam and Philadelphia: John Benjamins Publishing Company, 1995.

Schumacher, John N., editor. *Readings in Philippine Church History*. Quezon City: Loyola School of Theology, 1979.

Scott, William Henry. *Barangay: Sixteenth-Century Philippine Culture and Society*. Quezon City: Ateneo de Manila University Press, 1994.

Shorter, Aylward. "Inculturation, the Premise of Universality". In *A Universal Faith?: Peoples, Cultures, Religions, and the Christ*, edited by Catherine Cornille and Valeer Neckebrouck. Louvain: Peeters Press, W.B. Eerdmans, 1992.

Spunta, Marina. "The Facets of Italian Orality: An Overview of the Recent Debate". In *Orality and Literacy in Modern Italian Culture*, edited by Michael Caesar and Marina Spunta. London: Modern Humanities Research Association and Maney Publishing, 2006.

Tiongson, Jaime F. Laguna Copperplate Inscription: A New Interpretation Using Early Tagalog Dictionaries. Paper presented at the 8th International Conference on Philippine Studies at Quezon City, Philippine Social Science Center, 23–6 July 2008.

Wiegel, Katharine L. *Investing in Miracles: El Shaddai and the Transformation of Popular Catholicsm in the Philippines*. Honolulu: University of Hawaii Press, 2005.

Part II

INVENTING LANGUAGE TRADITIONS

5

THE TRANSLATION OF BUDDHISM FROM ASIA TO THE WEST

Shifting languages, adaptive logics, acculturations

Lionel Obadia

Buddhism can be read and heard in many languages. From sacred (ritual) languages such as Sanskrit and Pali, to *lingua francas* (such as Chinese), notwithstanding numerous vernacular tongues, over the course of two and a half millennia, the languages and cultures of Asia and, now, the Western world, have infused Buddhism with their particular colours and sonorities. Many voices also speak up in the name of Buddhism and produce discourses defending Buddhist "points of view" (on social issues or in the context of philosophical and political debates). Buddhism appears to resonate in a particularly acute way with the modern world. But while it has been adapted to Asian and Western languages, whether religious or profane, it resists thought simplification and is delineated into many shapes and discourses. So far, no global outlook on the history of Buddhism through its languages has been done, although many academic works feed into this direction. This chapter aims to go beyond this task and endeavours to examine the varied repertoire of ideas in the matrix of which Buddhism has been translated. It especially looks at the way current social, economic, and ecological issues allude and refer to key Buddhist concepts. As such, the chapter not only presents these new registers of meaning, but also questions the various relationships between Buddhism, language, and languages.

Language is not tongue – a Buddhist view on Buddhist linguistics

Buddhism is a religion for which it is fundamental to refer to texts, to the canon. As such, it is an exemplary scriptural religion. Buddhist scriptures are made of an ensemble of texts which, according to experts, amount to at least eighty-four thousand – counting only the most official literature. The even more multitudinous exegeses of the founding texts (the *Tripitaka*, or Buddhist canon), and the innumerable schools of thought, lineages, and traditions that were founded based on exegetic processes, with each lineage or school developing its own corpus and its own hermeneutic, should be taken into account. As a consequence, Buddhism's linguistic and textual content is even vaster than its original canon.

Before approaching Buddhism's repertoire of interpretation, one should remember that Buddhism, as a religious and philosophical doctrine, develops complex theories about the relationship between thought and language. Many Buddhist theories on language question the paradoxical nature of the relationship between speech/words and the world and, by extension, the nature of reality. In its most advanced philosophical form (the study is limited here to approaches from two major schools, *Madhyamyka* and *Yogacara*), Buddhist language theory states that words can never describe the ultimate transcendental experience, which is unfathomable, hence untranslatable. As a consequence, language remains bound to the "mundane" (*samsara*); to awaken (*bodhi*) means to overcome this condition and free the mind by the cessation of perception (*nirvana*). Nevertheless, through in-depth exegetic works, numerous discourses have been produced. They have mainly been focused on translating Buddhist ideas and doctrinal concepts into understandable words, into doctrines (Cabezon 1998: 5). As such, Buddhist theory of language tends to centre on *language* in its *textual* form and leave out the exploration of languages and of cultural repertoires of interpretation. As José I. Cabezon recalls: "From the outset we find tremendous interest in language on the part of Buddhists and indeed, if sources portray this accurately, on the part of the Buddha himself"; "the Buddhist attitude toward language was something that developed and became more sophisticated [. . .] with time. Arguably, this reaches its high point in the speculation of Buddhist scholasticism" (Cabezon 1994: 3).

Buddhist philosophy of language fluctuates between linguistic, semiotic, and pure semantic, which is interesting in itself. However, the other, more pragmatic and anthropological aspect will be explored here, by concentrating on the languages in which Buddhism has been historically communicated, especially looking at the recent phenomenon of Buddhism's transmission to the West. Because studies in Buddhism or about Buddhism on this topic have been very uneven and diverse, the issue of language turns out to be tricky.

First, it should be remembered that Buddhism is not a *culture*. It is a *religion*. It has spread extensively through mission and migration. And as can be observed in traditions which share the same nature (Islam, Christianity), languages and social ensembles do not overlap: Buddhism has been expressed in many languages, but there is no such thing as a "Buddhist language". Sanskrit acts in the same way for Buddhism as Latin for Christianity: it is a liturgical and learning language but is not used in daily life. And, very importantly, it is sacred. As such, its usage has been limited and circumscribed to specific contexts. Pali, its Sinhalese derivation, which was used to preserve the Buddha's sutras after his death, abides by the same phenomenon. These ancient languages have mainly served to codify the doctrine and have been used by priests in a liturgical context. Over time, the number of people who could understand it narrowed to a few thousand scholars and intellectuals. But historical episodes, although rare and limited, have brought it back to the historical scene. Sanskrit is also Hinduism's sacred language. Nationalist movements in India used it as an ideological resource for their identity claims. As India's "first", "original" language, it was brandished as the embodiment of the Indian "soul". Moreover, the Buddhist variety of Sanskrit and the founding of Pali partake on complex dynamics. As Ryūichi Abé points out, they are new languages and new scripts that were created *by* Buddhism; meanwhile, local languages simultaneously and reciprocally shaped Buddhism over the course of its development and diffusion (Abé 2005: 292).

From Buddha's muteness in Buddhist texts to the more simple speechless meditation in Japanese Zen schools (Sōtō or Rinzai), Buddhism is also a religion that knows how to handle silence. As such, it can be communicated through the absence of language, although background meaning is strongly present. Silence, whether the Buddha's in the sutras or the Rishi's in Zen tradition, is as loud as words here. On the other hand, Buddhism can also "exhaust" language by excessive usage and glossolalia. This is particularly visible in Tibetan currents. Through the influence of Tantras, *mantras*, which are phrases that do not make literal sense but are charged with codified meaning, are extensively used, generally through never-ending repetition. Language in *default* for one branch, in *excess* for the other: such is the spread of Buddhism's relationship to language. These very diverse aspects already point to the vast panorama of possibilities and possible analyses in the field of formal linguistics. It should also be noted that Buddhism has given rise to specific genres. The short aphorisms of haïkus and koans characterize Zen Buddhist schools. And its insertion into the repertoire of sagas, tales, and legends also facilitated its diffusion with popular classes. But two essential dimensions of this relationship are yet to be explored: first, that of the idioms and languages (Asian and Western) through which Buddhism spread and which applied acculturation processes to it; second, that of its repertoires of meaning, which provide the grounds for reinterpretation.

Buddhism in translation: history and processes

"Buddhism" can be described as a religious system, anchored in history and in societies. But the analysis here calls for another approach. For the purpose of this study, Buddhism will therefore be presented as a stock of symbolic resources, kept in the form of textual and oral traditions which have been shared, translated, and disseminated over the course of history. Taking Buddhism as a set of resources opens up two possibilities. First, it enables languages to be considered important factors in producing new and changing older shapes of Buddhism. And second, it brings to light the ways and modes in which specific aspects of Buddhism have been interpreted into various repertoires. The Buddha's doctrine, the *Dharma*, was formulated during councils and recorded by his disciples in Sanskrit, India's ancient language. Sanskrit was Buddhism's first *lingua franca*, and evolved to a specific incarnation through this influence: "Buddhicized" Sanskrit is different from classical Sanskrit, which is found in Hindu and Brahman texts. Depending on usage and media; on whether it was used for liturgy or for scriptural purposes; and, in the case of the latter, depending on the type of text, Buddhist Sanskrit further evolved to take on various shapes and forms (Lamotte 1976).

The spread of Buddhism followed several historical steps. Historians describe a first transmission route through the South in the third century BC. A second route opened northward from the first century AD. It then reached its maximum expansion to Asia in the fourteenth century. The Buddha encouraged missionary work. But migrations, contacts, and exchanges between civilizations were instrumental in its expansion. Traveling "translator monks" also supported its spread and played a large role in the (original, Sanskrit) Buddhist canon's being translated to Pali, Chinese, and Tibetan. The itinerant monks also fed the oral traditions of all the vernacular tongues encountered along the way (Bechert and Gombrich 1984).

Over the past two centuries, Buddhist traditions have left Asia and spread from their countries of origin, reaching new territories – i.e. Western lands – by means of migration, mission, and mass-media communication. As had been the case with previous diffusions, Buddhist ideas and practices have undergone a process of *translation*, shifting this time from Asian languages (Tibetan, Chinese, and Japanese) to Western ones (English, French, and German). The diffusion of Buddhism in the West since the nineteenth century offers a golden opportunity to observe in real time a new phase of translation and enables the observer to compare the effects and aspects of rupture and continuity with anterior phases. Needless to say that in terms of methods and available networks, the nineteenth and twentieth centuries are a far cry from what the world was like two thousand years ago. Societies have undergone technological and cultural globalization, and the means to communicate have multiplied (Tomlinson 1999; Waters 2001): press and the media, literature, cinema, television, the web, and all the rest constitute a new Babel, where ideas, texts, signs, and images mingle and mix in a mosaic of languages and registers.

As parts of society, religions are also subjected to the same processes and exchanges (Obadia 2010). No exception, Buddhism has had to adapt to the technocultural revolution which has impacted its environment and context, at the risk of altering its forms (Herschock 1999). On the Web, a simple search unearths millions of sites in Western languages dedicated to Buddhism or claiming to be Buddhist. But Asian languages are far from being used by all, and hypertexts are not necessarily written for religious purposes. Studies that have looked at Buddhism online have shown that the internet provides a double act of translation and diffusion (Prebish 1999a). Nevertheless, from the first Buddhist texts written on palm leaves and wood, to its unbound extension through modern media, a continuity can be observed: the transmission may change media, but its intentions, actors, and effects remain the same (Snelling 1987). Apart from the technological context and the linguistic extension of Buddhism's repertoires, nothing is new in this dispersion: associated with the technical supports of its time, and following the geographic circuits of its circulation, Buddhism and its ideas have always been communicated in other idioms. But with its worldwide spread in the twenty-first century, after – and it is key to keep this in mind – two centuries of Western diffusion (nineteenth and twentieth centuries), the stakes and challenges of translation reach a new dimension.

Buddhism in the West: beyond languages, conceptual circulation, and translation

The question of translation is raised, throughout history, in *linguistic* and *idiomatic* terms on one hand, and in *semantic* and *cultural* terms on the other. The linguistic issue – that of translating reality and of idiomatic variations – does not exhaust another issue, which is as fundamental: highly semantic repertoires of meaning point to the way Buddhist (scriptural) concepts have been reinterpreted, in bulk or as units. For the past twenty years, I have studied the diffusion of Buddhism in the West and its globalization. The analysis is therefore based not on tracing out and delineating previous diffusions in Asia, but on information gathered in that particular time and space, that of Buddhism in the West. The issue of *language* is articulated in different terms in Asia and in the West: Buddhist texts that had already been translated to Pali, Chinese, and Tibetan were subsequently translated into

Western languages (French, German, Spanish, and English for the more active orientalist milieu, but also into Eastern European languages). The history of the *linguistic* translation of Buddhist texts, as well as the *cultural* translation of Buddhism in culturally acceptable terms, is well-informed. At the very beginning, in the nineteenth century, there was no explicit strategy of translation from the many Asian languages in which Buddhist ideas had been transcribed, collected, conserved, and put in practice, to the handful of Western languages. A new generation of scholars, who would eventually become "Buddhologists", were at that time trying to understand the core notions of what was for most intellectuals a "strange" thought that looked like a religion without the distinctive features of one (Droit 1997).

Amongst the first attempts, one can note the efforts of Eugène Burnouf, credited as the forefather of Buddhist studies, and founder of the *Société Asiatique*, to translate Buddhist texts (in particular, the famous *Lotus Sutra*) from Sanskrit to French in 1852. British scholars were also active: Thomas Rhys Davids founded the Pali Text Society in 1881 and launched a vast programme of translation of text from Pali to English (and German, with Herman Oldenberg), Pali being the second main language in which Buddhism has been translated, in Ceylon. The "Belgian-French" school of Buddhology (according to Prebish 1999b), relying upon Sanskrit sources, also expanded the scope and the languages of Buddhist studies by including Chinese and Tibetan sources. In France, Jean-Pierre Abel-Rémusat and Stanislas Julien were instrumental in the exploration of Chinese sources, while in Italy, years later, Giuseppe Tucci worked actively on Tibetan sources – those two sources are themselves translations of the Sanskrit canon. The "Anglo-German" school (Prebish 1999b) remained faithful to the focus on Pali texts but eventually turned to Japanese sources, when Buddhist studies settled in the United States and when Paul Carus promoted the translation of the works of Buddhist Master D.T. Suzuki. In this context of blossoming of Buddhist translations in different national and linguistic contexts, Sanskrit was considered the main, because it was the most ancient, reference – and it still is today. The linguistic translation of Buddhism was conditional to the development of area studies and the corresponding linguistics: Buddhology, as such, was at the crossroads of Indology, Sinology, Japan studies, and Tibetology. As a consequence, the translation of Buddhist scriptures was touching upon a variety of idioms on both sides: the Asian sources and the Western targets. Faithful with the chronology of the texts themselves, Western scholars considered Sanskrit (and secondly Pali) a reliable language for the translation of Buddhism, since it was the "authentic" first tongue of Buddhism itself. Nowadays, Sanskrit references and notions are still seen as the "linguistic yardstick" from which the (Westernized or modernized) ideas and images of Buddhism are framed – and when Buddhism is quoted as a cultural, spiritual, or philosophical touchstone, Sanskrit terms are always used, even if they are mixed with non-Asian languages. Sanskrit retains the authority of its antiquity even though Buddhism has travelled to the West through other languages (Chinese, Tibetan) and finally ended up in Western idioms.

These new texts provided information which quickly overflowed the boundaries of specialized groups. They spread to larger intellectual circles in the nineteenth century and, as technology developed, came to gradually invest and develop in other cultural sectors (literature, press, and audio-visual media) (Baumann 1997; Lopez 1998). The more they spread, the simpler the Buddhist themes and words became, losing their connection to the complex systems of ideas (doctrines and philosophies) from which they took their meaning and in which they made sense. What all these translations of Buddhism share is that they have kept a basic catalogue of terms – in Sanskrit and Pali – which make up the key concepts (substantives and verbs) that are integrated into the interpretation process: *nirvana*, *Buddha*, *dhukkha*, *sunyata*, and *sangha* are common terms found in all interpretation attempts. Through this filter, only a handful of key concepts which are at the lexical and theoretical heart of Buddhism's doctrines (the "pivots" of Buddhism) have been preserved, spread, made known to, and appropriated by vast numbers of people in the West, taking on new meanings which are more in line with modern societies' cultural models and (secular) ideals.

However, after a phase of idiomatic and semantic loss, with the installation of practising communities made of migrant descent on one hand and converts on the other, and the implantation of monastic institutions, keeping in mind the complex soteriological and philosophical systems, the ties to "tradition" found new strength. This was especially true in the case of the training of Western monks and ascetic disciples: as they learned the sacred languages of whichever tradition their school belonged to (Tibetan in Mahayana, Japanese in Zen, etc.), an elite perpetuated the more orthodox forms of Buddhism, resisting the numerous "consumers" of a free-for-all Buddhism. In that way, words and meaning oscillated between concentration and dispersion, between lexical and semantic regulation and free interpretation. Even in practice, the issue of maintaining traditional languages for new converts was raised. What was at stake was the acclimation process: should Buddhism, as Christianity did, go through an *aggiornamento* and forgo its ancient Asian languages to select the idioms of its new chosen lands?

Let's leave the theme of linguistic translation to scholastic specialists: this chapter will look at semantic repertoires that have provided translation matrixes to Buddhism as it spread beyond Asian borders in modern times. In parallel to the idiomatic translation, Buddhism has indeed undertaken cultural translations. The reinterpretation of Buddhist themes and ideas through a series of specific repertoires corresponds to mainstream cultural frameworks at certain times in Western history, springing from prominent systems of knowledge (physics, psychology, etc.) and, more recently, from more ideologically oriented frameworks and concerns (feminism, ecology, etc.). The second part of this chapter therefore focuses on some of the most visible repertoires through which Buddhism is translated and interpreted: Christianity, scientism, psychologism, feminism, atheism, ecologism, and economism. The repertoire could have been declined in a more precise and detailed way by including humanism, or postmodern philosophy, for example (Obadia 2007). But for the sake of pertinence and available space, the emphasis will be exclusively on these seven matrixes.

Christianity's reverse mirror: religion, atheism, and scientism

The very first translation and intelligibility matrix of Buddhism is Christianity: when far-Eastern religious traditions were discovered in the nineteenth century, Christianity was the dominant religion in the West (undergoing the process of secularization). The Buddhist tradition was long ignored, as it was confused with other religions from the Indian sub-continent. Quite logically, once discovered, it was compared with what was (and *still* is) the local reference system for religious sciences, as provided by Christianity. Native Christian concepts of God, faith, and religious and spiritual goals for rites and liturgy quickly led nineteenth-century Western thinkers, philosophers, and theologians to consider Buddhism with caution: with no apparent God, no transcendence, and a religious model that appeared to be reduced to personal asceticism, Buddhism lacked the recognizable traits of religion. The "Buddhist issue" in nineteenth-century Europe provided fertile grounds for debate in all sorts of intellectual milieu, centring on questions around the religiousness (or lack thereof) of Buddhism and its very nature.

What happened in the nineteenth century is key to understanding the modes of translation of Buddhism, their effects, and their contexts. At that time, a major phenomenon emerged: that of the evolution from a *religiologic* approach to Buddhism to a *scientist* outlook. Catholic and Protestant theological terms provided comparison scales. Simultaneously, another major system of knowledge – science – which, paradoxically, had emerged during the Romantic period, tackled Buddhism's key concepts (*nirvana*, *samara*, *bodhi*, *Bouddha*, etc.). This explains why Buddhist themes were concomitantly translated in terms that related, and are still considered as relating, to the scientific register (Wallace 1998).

As a semantic category, "science" is very plastic: many fields of knowledge share the "scientism" umbrella, although it was originally meant to represent only "hard" or "exact" sciences, such as physics. The theosophist Henry S. Olcott based his analysis on this aspect when, in his now renowned *Buddhist Catechism* (1881), he reckoned that Buddhism was "scientific in its outlook". His book fathered the confusion between what pertains to comparing scientific fields (such as astrophysics, with the dialogue between Matthieu Ricard and Trinh Xuan Thuan 2004) and what can be more accurately described as apologia. The latter takes on two forms: the apologia of science through the lens of Buddhism, as with Einstein (already), on one hand; and an apologia of Buddhism with references to science, which is much more common, on the other, as evidenced by the numerous books, written by Buddhists or Buddhist partisans, which include the word "science" in their title.

Atheism

Atheism, being the essence of the two-and-a-half-millennia-old tradition, is one of the most recurring reinterpretation repertoires (in the West). In modern Western societies and, as a result of the globalization of ideas, in Westernized/modernized Buddhist Asian societies as well, Buddhism is often brandished as a type of atheism and therefore signalled as remaining compatible with secularized modern societies. But, in fact, "Buddhist atheism" appears to be a Western construction linked to the examination of ancient texts. This probably started when orientalist Herman Oldenberg declared Buddhism to be a "Godless religion" (1881). This particular reinterpretation matrix is based on the divinities', or *devas*', inferior status as compared to the Buddha's powers. Emile Durkheim ratified this point of view (1912). And, since then, many generations of commentators on Buddhism have followed suit.

More recently, it has become more common to separate, on one hand, what can be considered doctrinal atheism – which, in the socio-historical world, has never existed as a concept apart from the contexts of theistic belief systems (Spiro 1972) – and, on the other, ethical or ideological atheism, which springs from secularization in the West. Since then, the lexical reach of the atheist grid of interpretation for Buddhism has grown and expanded: not only do many Westerners associate Buddhism with atheism, based on the questionable grounds that it is not a religion; but many scholars also rally to this theory, as do some Buddhists, such as Stephen Batchelor (1998), who choose to defend the idea that Buddhism does not need and should be "freed" from gods and beliefs.

Psychology

Psychology represents one of the most called-upon registers for the translation of Buddhism in the West. Since the beginning of the nineteenth century, under the impulse of Eugene Burnouf, who founded the first Institute of Buddhist Studies in the early nineteenth century, the focus has been on understanding the more rational aspects of Buddhism from its texts and through its texts. It is well-known that the Buddhist theory of mind, on which is based the soteriology of recognizing the impermanence of the world (*anitya*) and the relative aspect of reality (*samvrti-satya*), and its theory on mental processes (*Skandha*) leading to their extinction (*nirvana*), have been widely discussed by philosophers of the mind and the pioneers of nascent psychology.

Within the field of psychology, several sectors and currents found fertile grounds in the confrontation with Buddhist concepts. Buddhism caught the interest of Nietzsche and Schopenhauer, insofar as they could incorporate it into their respective philosophies of existence. As such, Buddhism went through a double translation process, both idiomatic and semantic, into German; and the idea of vacuity (*shunyata*) became associated with nihilism in European philosophy. The founding father of psychoanalysis, Sigmund Freud, also found inspiration in Buddhism, by exploring the concept of *nirvana* and integrating it into his own theory on drives. With Carl G. Jung, psychoanalysis took on a more "spiritualist" orientation. Jung opened up more space for symbolic approaches, but also considered Buddhism as embodying an "Eastern psyche", which was seen as fundamentally introspective. Both psychologists interpreted the Eastern tradition through their Western eyes. They were at the onset of the dissociation between a cold, neutral, distanced, and objective approach to Buddhism in psychology (Freud) and the more emotional, intimate, personal approach developed by Jung. Through the collusion between psychology and oriental traditions, Jung tried to shed light on the human mind's mechanisms and, beyond, on the psyche's universal structures. This dissociation, however, then became more blurry, as the border between scientific studies and spiritual adherence to Buddhism dissolved.

Psychological interpretations of Buddhism continued with the dialogue on Zen between Fromm and Suzuki in the 1960s. In the 1970s, the first clinical studies on the effects of Tibetan meditation took place (in Boulder, Colorado): for the first time, electronic devices were used to obtain scientific measures. The available tools have subsequently developed, and so have the theories, coming from discoveries in cognitive sciences and neurosciences. Interestingly, many well-known Tibetan monks have participated and participate in such studies.

As this phenomenon moved forward in time, the "psychologization" process of Buddhism became more manifest. It is especially visible with the organization of MindScience days in the United States. These are set up with the goal of opening a space for systems of ideas and concepts (Buddhism and Western psychology) to be debated and confronted, through the organizers and participants' shared desire to explore the deep mechanisms of thought. This has led to the theory, however questionable, of identity, which claims that Buddhism anticipated modern psychology. Participants' conclusions tend to point towards a shared, universal nature of mind, which is expressed in different languages and in different cultures. In this linguistic and semantic confrontation, psychology nevertheless retains the dominant position and provides the grid for interpretation. It is partially responsible for the current scientific legitimization of Buddhism and finds ideological backing and justification in the quasi-scientific rationality that the Buddha's religion seems to deploy when analysing the way the mind operates. This interpretation, however, reaches its limits when repertoires are brought back to their contexts. And some scholars point out that more than two thousand years separate the birth of Buddhism and that of psychology, bringing some relativity to the debates.

Feminism and issues in gender

Simultaneously, counter-culture movements, feminism, and issues in gender also provide a major interpretation register, mostly in North America. Monastic institutions from South-East and Northern Asia (Theravada and Mahayana Buddhism, respectively) are settling in Western lands; individuals from local communities, which carry the value systems of sexual equality, are being converted; and Asian migrant descendants acculturate in the West, integrating such equalitarian systems: Buddhism has served as a tool to contest (Western) male domination, based on its universal principles (especially visible in Mahayana) of the Bodhisattva's salvation beyond social, racial, or sexual differences. This was particularly brandished by the ever-more-numerous women who became fervent disciples and, sometimes, even renowned masters of Buddhism. Within "native", Asian Buddhist communities, new generations have come to contest even more loudly the fact that monastic power is based on gender (Cadge 2004).

The lexicon of feminism and gender is therefore rooted in Buddhism seen as "androcentric", the slight of which it is called upon to shed. The rhetoric of sexual equality was even brought further, to Queer Buddhism, founded in the United States: a gay-friendly movement of converts which does not seem to be linked to the fact that many Western Buddhist masters themselves admitted to their homosexuality (Leyland 1998). In Buddhism, sedition against the established order is a strong aspect, especially as it takes into account the social perspective of castes, kings, and rulers being equally subject to death and spiritual realization. Shakyamuni Buddha was far from the revolutionary activist as whom he is sometimes portrayed. But when it comes to sexual liberation, the *gendered* repertoire dresses his mythical figure and historical action with the cloak of a pioneer. However, traditional representatives from religious institutions, even the more progressive ones, such as the Dalai Lamas, barely abide by such reading.

Ecology

Cultural translations of Buddhist themes and ideas in the West follow the trends of dominant or culturally important ideologies. One of the more recent, and topical, frameworks of thought and terminological repertoire comes from ecology. The massive destruction of the natural environment, generalized industrialization and its numerous immediate environmental costs (soil and air pollution, depletion/exhaustion of resources, mass waste production), the sensation of ecological risk and its many variations expressing crises in similar terms (productivity, wealth distribution, biotope transformation, and, more recently, recognition of climate change due to human activity), have forced religions to participate in the debates and prove that they understand elements, or even whole chapters, of ecological theory. This justifies their inclusion in a discussion which was previously reserved to states and large companies.

More and faster than other religions, Buddhism found itself on the front line of such debates. Although based on arguments found in ancient texts, the idea that there is such a thing as "Green Buddhism" or Eco-Buddhism is a recent phenomenon aiming to match some aspects of Buddhist philosophy, especially the notions of interdependence (*pratityasamutpada*) and altruism (*karuna*), with current concerns for the planet and its inhabitants as they are expressed by ecology (Johnston 2010).

The question of translation and lexical and semantic variations is particularly stark here: was Buddhism originally "ecological"? If so, were these aspects revealed only when the modern world started worrying about these issues? Or has modern ecology colonized Buddhist approaches to the point of bringing about a translation in the words and terms of modern, secular, Western ideology (Obadia 2012)? In the latter case, ecological doctrines that belong to Buddhist ontology have been translated into more current terms. And notions of "interdependence", respect for the environment, non-destruction of life and, by extension, of the world, and the idea of cosmos, have been coined. Or, looking at it from the latter perspective, they have been reinvented after going through the lens of modern ecology – that is, *naturalist* ecology (which deals with preserving biotic and abiotic environments), whereas traditional ecology in Buddhist countries is social (i.e. centred around the notions of purity, allowing the world to follow its rightful course, etc.) and very bound by social relations.

Although they divide the Buddhist milieu between sceptics, on one hand, and those who are convinced it is true and useful, on the other, modern shapes of Buddhist ecology are now influenced by new standards and take a naturalist slant which was previously unknown. As a paradoxical consequence of such phenomena, modern ecology is now being coloured with Buddhist references (which have been modernized and secularized) to justify preserving nature and the natural climate. In this way, words and ideas come full circle.

Buddhism, economics, and economy

One of the most recent repertoires which mobilize Buddhism to make sense of the modern world is economics: both *economy*, that of the dynamics at play in financial markets and consumption circuits, and *theoretical and scientific economics*, that of universities and expert circles. The great living Buddhist masters of the twentieth and twenty-first centuries, especially those who are bound by the logics of internationalization by representing globalized figures of religion (such as the Dalai Lama in Tibetan Buddhism or Thich Nhat Hanh in Zen), have strongly criticized modern capitalism: they have accused it of alienating the world through its logics (turning everything into marketable merchandise), its thoughts (instrumentation), and its practices (all-out consumerism). The translation of Buddhism into economic terms may appear particularly paradoxical: the Buddha's founding ideas were based on withdrawal from social and productive life and on monastic vows of poverty (*bhikkhu*). Modernist currents specifically position themselves in contesting the expansion and impact of capitalist market economy. But Buddhism's encounter with economics follows the same path as its meeting with psychology: the opposition of repertoires (religious versus profane, spiritual versus material, metaphysical versus mundane) gives birth to an unlikely, yet very significant fusion, which is now termed *Buddhist economics.*

This movement was initiated by German economist Ernst Schumacher: in the 1960s and again, at the beginning of the 1970s, he published a short book in which he suggested that Buddhist scriptural principles be applied to modern economy to make it more humanist and respectful of humans (workers as well as consumers). However utopian, the far-fetched fusion of the two systems is still limited in its scope and reach. But it has created a space for debate on the possible complementarity and opportunities to

"Buddhicize" economy. Injecting Buddhist glossaries and discourses into economy and economics has, again, generated the reinterpretation of theoretical economics through the Buddhist lens. In turn, this has influenced concrete material economies: in various corners of the world, both in the West and in Asia, some people attempt to apply Buddhist "recipes" to productive economy, to the corporate world, and to financial markets.

Some years later, in 1982, and on another register, French economist Serge-Christophe Kolm painted Buddhism as embodying the rationality of *homo economicus*, opening up a whole new chapter for Buddhist economics. He tried not so much to apply Buddhist concepts to effective economic life as to conceptualize religious behaviour into models that could be transposed to economic theories. His attempt to bring together two unlikely systems took place even before the global diffusion process at play today came to impact political and monetary economy – although it should be remembered that this influence remains quite marginal.

Translation stakes, adaptation at stake

These few examples show how complex the various levels of translation are when it comes to Buddhism. The sacred idioms of Buddhism have gone through vicissitudes when transferred to Western languages, and the repertoires of religious ontology (Buddhist dogmas that offer an explanation of the world) have evolved in unexpected directions upon their encounter with modern and secularized ideologies. With no direct, immediate lexical correspondence, all semantic mergers become possible, depending on the matrix of interpretation that is being used and on the freedom found when undertaking the translation process. In the case of Western Buddhism, the issue of literally translating doctrinal and liturgical content raised complex questions, highlighting the intricate background processes that affect it.

Over the course of (recent) history, Buddhist ideas in the West have been subjected to a process that can be termed *inculturation*. This word refers to the Christian missionaries' strategy: by adapting their speech and practices to local languages and cultural customs, they could more easily alter them after conversion, beyond the usual *acculturation* processes of tradition metamorphoses, and independently from actors and their intentions. The corpus of translation obeys two different dynamics: (a) interest in learning and knowledge and (b) proselyte motivations. This aspect appears obvious and implicit through all translations of Buddhist themes and concepts over the past two centuries. For that reason, the analysis is enriched if it goes beyond the level of dynamics and *linguistic* exchanges through which Buddhism circulates and changes meaning and sometimes even words. It calls for an exploration of the stakes and *pragmatic* challenges of the strategies of language and discursive choices which are adopted by the various social and historical actors involved in translating Buddhism. This enables us to explore what is at stake in translation, rather than simply describing the modalities and contexts of translation.

In the background are the various forms taken on by the translations, moulded by all the ideological systems which seek to adjust Buddhist concepts to the great, mainstream ideas of certain times in the host contexts: it was "atheist" in nineteenth-century secularizing Europe, it was "scientific" at the beginning of the twentieth century, it is "ecological" in response to today's concerns on the global scale, and it was "gendered" a decade ago to attempt to answer to social problems encountered by gender minorities.

Buddhism's new poetics rally with enunciation politics: in a polylogical world (Kavolis 1991) where religions communicate through both registers of poetics (aesthetic enunciations) and political (instrumental and strategic enunciations) (Kong 2001), Buddhist repertoires and the Western ideologies they have met and impregnated and which, in turn, have reinvented it, participate in a complex web of translations and translations of translations. These point to the very adaptable quality of lexicons, idioms, and semantics in contact. But the interpenetration also generates many resisting knots. It is difficult to establish which interpretation matrix actually manages to dominate: is Buddhism being psychologized/ecologized/economized? Or are psychology, economics, and ecology being Buddhicized? The debate is open, and this is not the place to attempt a definite answer.

Ryūichi Abé beautifully stated that "Buddhism is arguably the most loquacious religion". He justified his theoretical catchphrase by the fact that "not only are its teachings available in hundreds and thousands of volumes, but its missionaries have worked avidly for centuries to render these scriptures into many Asian languages and wrote numerous commentaries on them in an even greater number of languages". He added that "Buddhism is, in short, a mass producer of sacred words" (Abé 2005: 291–2). Buddhism's garrulousness, which was already its hallmark in Asia, found new grounds to spread in the West, within the context of globalization. There, it was met with the talkative languages of faith and non-faith, of world knowledge and human knowledge, that the West in turn had produced, apparently inspired by Buddhism.

Conclusion

This short and very general panorama of the modes by which ideas, concepts, and themes and, by extension, of Buddhist practices (although this dimension has barely been touched upon here), are translated and reinterpreted is based on fairly limited examples. But the analysis nevertheless brings forward several logics at play in the relationship between Buddhism and language/languages. History shows that, on one hand, the transformation of Buddhist theological elements as a stock of symbolic resources was transmitted from one generation to the next in Asian societies and spread through mission and migration, adapted to the shape applied to them by social and cultural exegetic institutions (the monastic order), which created innovation in semantics (meaning) within the same lexicality (corpus).

This explains the multitude of Buddhist schools of thought, from Shakyamuni to our time: continuities, schisms, and ruptures have not broken the shared ground of referring to the same texts. If Buddhism is translated into different languages – Pali, Chinese, and Tibetan – what is at stake here is language, and not idiom: whatever tongue is chosen, Buddhist exegesis aims to extract the meaning of the key concepts Buddha formulated (*dharma*, *sangha*, *nirvana*, *bodhi*, etc.). This basic dimension pertains to what José I. Cabezon called *scholasticism*, which he defined as the way in which learning, repeating, and understanding Buddhist doctrines within a specific social context is bound by institutional vocabulary. Another dimension is that of idiomatic variations within vernacular languages that transported Asian Buddhism throughout Asia, then to Europe, and, from there, to the whole Western world, up to the point when the ancient tradition went global. Finally, the third, and most important, dimension touched upon here is the ideological filters applied to Buddhism: Buddhist ideas have spread beyond the control of monastic institutions, appropriated by social representatives who are neither monks nor masters. This process allows much leeway to reinterpret key concepts through dominant thought systems, severing the link to their original context: rationalism, scientism, humanism, ecologism, economism, secularism, and feminism are some of the many semantic frameworks that sediment into new levels of reinterpretation and translation for Buddhism – which had already undergone major transformations before it reached the West.

Most Western scholars who have studied Buddhism have highlighted the essential part played by "free" translation within the context of these numerous ideological frameworks. But they sometimes forget that very traditional forms of language, discourses, and word usage had already been reinvented within the Buddhist mosaic of institutions. These institutions have indeed managed to preserve the regulatory mechanisms towards ideas and speech – as is the case with any monasticism – despite the significant changes in their cultural and ideological environments, through the Ages. Further, in the case of Buddhism, the non-Buddhist,

Western, secular and modern themes and repertoires of interpretation of this ancient Asian spirituality have strategically been used by monks and religious actors in order to adapt their tradition to a Western, modern and secular environment (Obadia, 2007). Either way, all these systems of ideas have contributed eclectic interpretations of Buddhist ideas and practices, in a series of layers of cultural references. In parallel, they have participated in the reception and settlement of Buddhist traditions in Western countries.

To conclude, this chapter sought to outline the complex translation processes at play in the context of Buddhism's modern Westward expansion. It emphasized the tension between idiomatic and cultural frameworks in the diffusion of Buddhist ideas and analysed the role of languages in the diffusion, adaptation, and transformation of Asian religious practices and institutions in the secular West. As a consequence, Buddhism has, in some ways, been "coloured" by modern Western societies' cultural frameworks, ideologies, and ideals, and it has evolved accordingly – at least on a certain level.

Indeed, in practice, Western Buddhism is torn between two conflicting trends: modernism and traditionalism. The former attempts to "universalize" Buddhism by renouncing "old-fashioned" rituals and sacred languages: on the opposite end of the spectrum, the latter is driven towards a "return" to classical literature and orthodox liturgy, which positions it as Buddhism's "ethnic" centre in the West. Asian sacred languages are therefore key to understanding the driving forces behind – and resistance to – change in Western Asian and "White" Buddhist communities. Moreover, transformations in Buddhist traditions can be measured far beyond the empirically observable dynamics of continuity and change: they are also visible in linguistic ritual and training customs. This also raises epistemological issues about the ways in which new "cultural tongues" impart their linguistic and cultural styles to Buddhism and reshape in a deep way the beliefs, values, institutions, and practices of the 2,500-year-old Asian tradition.

Has Buddhism been "psychologized", "greened", and "gendered" because it has become a topical, fashionable, and trendy item in psychology, ecology, and feminism? The answer is complex. On the basis of empirical evidence collected in France and the rest of Europe, and on a comparative review of recent literature from the United States and other countries, this chapter specifically attempted to examine the various epistemological positions and effective roles of languages, idioms, and tongues in the diffusion and acculturation of Buddhist traditions, as exemplified by Tibetan, Zen, and Theravada schools.

References

Abé, Ryūichi. "Word". In *Critical Terms for the Study of Buddhism*, edited by Donald S. Lopez, 291–310. Chicago: University of Chicago Press, 2005.

Batchelor, Stephen. *Buddhism without Beliefs*. New York: Riverhead Books, 1997.

Baumann, Martin. "The Dharma Has Come West: A Survey of Recent Studies and Sources", *Critical Review of Books in Religion* 10 (1997): 1–14.

Bechert, Heinz, and Gombrich, Richard, editors. *The World of Buddhism, Buddhist Monks and Nuns in Society and Culture*. London: Thames & Hudson, 1984.

Cabezon, José I., editor. *Scholasticism Cross-Cultural and Comparative Perspectives*. Albany: State University of New York Press, 1998.

Cabezon, José I. *Buddhism and Language. A Study of Indo-Tibetan Scholasticism*. Albany: State University of New York Press, 1994.

Cadge, Wendy. "Gendered Religious Organizations. The Case of Theravada Buddhism in America", *Gender and Society* 18, no. 6 (2004): 777–93.

Droit, Roger-Pol. *Le culte du néant. Les philosophes et le bouddhisme*. Paris: Le Seuil, 1997.

Hershock, Peter D. *Reinventing the Wheel: A Buddhist Response to the Information Age*. Albany: State University of New York Press, 1999.

Johnston, Lucas. 2006. "The 'Nature' of Buddhism: A Survey of Relevant Literature and Themes", *Worldviews: Environment, Culture, Religion* 10, no. 1 (2006): 69–99.

Kavolis, Vytautas. "Nationalism, Modernization, and the Polylogue of Civilizations", *Comparative Civilizations Review* 2, no. 5 (1991): 124–43.

Kolm Serge-Christophe. *Le Bonheur-liberté, bouddhisme profond et modernité*. Paris, PUF, 1982.

Kong, Lily. "Mapping 'new' Geographies of Religion: Politics and Poetics in Modernity", *Progress in Human Geography* 25, no. 2 (2001): 211–33.

Lamotte, Etienne. *Histoire du bouddhisme indien. Des origines à l'ère saka*. Louvain-la-Neuvre (Belgium): Publications de l'Institut Orientaliste de Louvain, 1976.

Lopez, Donald S. *Prisoners of Shangri-La. Tibetan Buddhism and the West*. Chicago and London: University of Chicago Press, 1998.

Obadia, Lionel. *Le bouddhisme en Occident*. Paris: La Découverte, 2007.

Obadia, Lionel. "Globalization and the Sociology of Religion". In *The New Companion for the Sociology of Religion*, edited by Bryan Turner, 477–97. Oxford: Blackwell, 2010.

Obadia, Lionel. "Is Buddhism Like a Hamburger? Buddhism and the Market Economy in a Globalized World". In *Research in Economic Anthropology*, Volume 31, 99–121. Amsterdam, Lausanne, and New York: Emerald, 2011a.

Obadia, Lionel. "Political Ecology and Buddhism: An Ambivalent Relationship", *International Social Science Journal*, special issue: "Political Ecology: Religion, Myth, Belief" 205/206 (2011b): 313–23.

Prebish, Charles S. *Luminous Passage: The Practice and Study of Buddhism in America*. Berkeley: University of California Press, 1999a.

Prebish, Charles S. "The Academic Study of Buddhism in America: A Silent Sangha." In *American Buddhism: Methods and Findings in Recent Scholarship*, edited by Duncan Ryuken Williams and Christopher S. Queen. Richmond: Curzon Press, 1999b.

Richard, Matthieu, and Trinh Xuan. *The Quantum and the Lotus*. New York: Three Rivers Press, 2001.

Snelling, John. *The Buddhist Handbook: A Complete Guide to Buddhist Teachings and Practices*. London, Sydney, and Auckland: Rider, 1987.

Spiro, Melford E. "Religion: Problems of Definition and Explanation". In *Anthropological Approaches to the Study of Religion*, edited by Banton Michaal, 85–126. London: Tavistock Publications, 1971.

Tomlinson, John. *Globalization and Culture*. Cambridge: Polity Press, 1999.

Wallace, B. Allan. *Buddhism and Science: Breaking New Ground*. New York: Columbia University Press, 2003.

Waters, Malcolm. *Globalization*. 2nd edition. London: Routledge, 2001.

6

A LANGUAGE 'CLEARLY UNDERSTANDED OF THE PEOPLE'

The construction of an Anglo-Catholic linguistic identity, 1850–2015

Amanda J. Haste

When the English king Henry VIII broke with Rome and created the Church of England in the sixteenth century, it was vernacular English which was chosen as the ritual language of the new Church. This meant underlining the break with Catholicism by rejecting the mystery of the Latin of the Catholic service books in favour of a tongue "clearly understanded of the people" in Archbishop Cranmer's new *Book of Common Prayer*. At the same time, Henry VIII also dissolved – and often destroyed – the Catholic monasteries, mainly for political and economic reasons, meaning that monastic life was outlawed and English monks and nuns were forced either to abandon their religious vocation or to migrate to (Catholic) mainland Europe.[1]

Three hundred years later, in the 1840s, the revival of the Anglican Church which was spearheaded by the 'high church' Oxford Movement led to the foundation of the first monastic communities in the Anglican Church. Over the next 150 years, some 200 Anglican monastic communities were founded, and more are still being founded today.[2] However, although the English of the *Book of Common Prayer* and subsequent rites has always been and remains the ritual language for secular Anglicans, for these monastic communities the choice has not been so simple. As John Edwards said in *Language and Identity*, "history reveals that, in many parts of the world at many different times, powerful and consequential connections have existed between particular languages and particular religions" (Edwards 2009: 100) and, until the mid-twentieth century, the language of the Roman Catholic Church in Europe was Latin. For those attempting to revive monastic life within the Anglican

Church, this meant that Latin, as the "sacred language" of Roman Catholicism (Joseph 2004: 100) thus acted as a "signifying medium" (Asad 1993: 131) of Catholicism, whereas the vernacular was symbolic of the Englishness of the Anglican Church. This research outlines the practical challenges faced by the early Anglican communities in their search for a musico-textual framework for their daily lives of prayer and worship, working from both Latin and English sources, and describes the processes involved as the communities developed their ritual repertories from the mid-nineteenth to mid-twentieth century; the story is brought up to date by touching on recent (2013) developments in the Anglican religious communities (ARC).

The second half of the paper provides an analysis of the changing role of Latin and English as respective signifiers of religion and nationhood and the role of language in current Anglican identity. It seeks to explain the communities' dilemma in a sociolinguistic context by drawing on ethnographic data gathered from present-day Anglican religious,[3] contextualized within the existing literature in linguistics and culture. The literature covers several related disciplines and includes the work on language and identity of John E. Joseph (2004) and John Edwards (2009); on ritual and performance by Talal Asad (1993); on issues in translation by George Steiner (1993), Nigel Armstrong (2005), and David Bellos (2011); and on religious, cultural, and national identity by Stuart Hall and Paul Du Gay (1996), Benedict Anderson (2006), and Eric Hobsbawm and Terence Ranger (2012).

Historical background

The revival of Anglican religious life was born out of the Oxford Movement's mid-nineteenth-century drive to affirm the Catholic heritage of the Anglican Church. Between the 1840s and the 1960s, ninety convents and twenty-nine monasteries were established in England, and in America a parallel high-church movement led to the foundation of forty-two convents and twenty-two monasteries.[4] While not all have survived, today there are some ninety Anglican communities, plus several hundred mission houses worldwide, although this chapter necessarily focuses only on native anglophone mother houses. The majority of the early communities combined active work with a life of prayer and worship, founding schools and hospitals, and providing social care in Britain and the Americas, but alongside their undoubted contribution to the social fabric of the modern world, they also felt the need to align themselves with their monastic heritage. Many communities (such as the Benedictines of West Malling, Kent) acquired the ruins of pre-Reformation monasteries and convents, while others, guided by luminaries of the Oxford Movement and the Cambridge Ecclesiological Society, designed new buildings. This resulted in imposing Gothic Revival convents and monasteries such as St Mary's Convent at Wantage (Oxfordshire) and All Hallows, Ditchingham (Norfolk) which offer us a clear physical indication of the communities' deeply felt need to "establish continuity with a suitable historic past" (Hobsbawm and Ranger 2012: 1). Each community, as well as building a new home, needed to create a liturgy and Office, and it is here that their need to identify with their monastic past brought them face-to-face with the problem of linguistic identity as they sought to create a brand-new, uniquely Anglican monastic ritual.

Monastic life is centred on the Daily Office – the full eight-fold Office consists of lauds, prime, terce, sext, none, vespers and compline, and the night office – but there was of course no model for English monasticism other than the pre-Reformation monastic rite. So the communities were faced with a choice between the Latin of the Roman Catholic monastic Office and the vernacular English of Cranmer's *Book of Common Prayer*, the latter being designed for secular use, even though its source material was the Benedictine Daily Office. Not only would their choices make an unequivocal statement about their theology and thus their allegiance, but also they would carry an additional layer of meaning: that of nationhood.

Born out of the high church Anglo-Catholic movement, the communities had a strong sense of their Catholic roots, but they were also conscious of belonging to the Church of England and, as Anglicans, were by definition not Catholic. [I well remember the insistence of our priest during my confirmation classes in 1970 that the description of the Church of England in the Creed as "catholic and apostolic" meant 'catholic' as in 'universal,' NOT Roman Catholic with a big 'C', *because we are most definitely NOT Catholics*!!" – my emphasis.] Added to this, the Church of England is *the* official faith of the United Kingdom, with the reigning monarch styled 'Defender of the [English] faith' and Catholics having suffered vilification and being effectively excluded from public life from the seventeenth century to the early nineteenth century.[5] The Anglo-Catholic movement, with its accompanying revival of fancy 'popish' ritual outlawed since the Reformation, had revived the old fear of 'papistry', meaning that the early communities were born into a period of intense anti-Catholic feeling, so they needed to establish boundaries between secular Anglicanism and (Catholic) monastic ritual in order to construct their own 'otherness' in a way that would be acceptable in the Church of England.

Their choice of either Roman resources in Latin, or Prayer Book texts in English, was therefore an important means by which they could create a 'constructed other': monastic rather than secular; Anglican, not Roman Catholic. As Philogene argues, "[i]n setting such a boundary with this constructed Other, one is defining oneself by defining who one is not" (Philogene 2007: 34). But these boundaries were very fluid, and they varied considerably between communities, with some identifying very closely with

Rome and others determined to carve out a specifically Anglican identity. This impacted directly on their choices. Some communities, particularly those known as 'Anglo-Papalists', whose aim was reunion with the Roman Catholic Church, adopted the Roman Rite in Latin from the beginning, and this immediately provided them with all the texts and music they would need for a full monastic Office. They could also feel the weight of history in the rhythm and shape of this mystical and sacred language, and maybe a certain exoticism which could not be served by the vernacular.

But for most of the new communities it was the beauty and poetry of Cranmer's vernacular prose, which had by then permeated English life, which they adopted. The Prayer Book services of Holy Communion, Mattins, and Evensong were themselves a conflation of the Catholic Breviary, but as the burgeoning convents and monasteries aspired to anything up to a seven-fold Office, these three reduced services were not sufficient. So these 'Prayer Book' communities still needed texts for the Daily Office, and they – or, more often, their scholarly founder fathers, who were mainly Oxbridge-educated clergymen – set about producing breviaries, which were English translations of the Latin.[6] In 1858 the need for a comprehensive Anglican Office Book was fulfilled with the publication of *The Day Hours of the Church of England*, newly translated and arranged according to the Prayer Book and the Authorized translation of the Bible. The nineteenth-century versions of the *Day Hours* were scrupulous in their observance of the *Book of Common Prayer* and the Authorized Bible; the book seems to have been the most popular version of the Breviary Offices among Anglicans, and it became probably the most widely used among the sisterhoods.

This work had to be approached sensitively: even those communities whose ritual practices were based on the *Book of Common Prayer* were treated with suspicion by the public and often regarded as Roman Catholics in disguise; they soon became known as 'Prayer Book Catholics'. Edward Pusey, who founded the first Anglican sisterhood, was all too aware of this as he prepared a breviary for the sisters' use. Keen to avoid the possibility of the sisters' desertion to the Roman Church, he omitted any matter regarded as incompatible with the formularies of the Church of England, including the Latin form of Confession and the comprehensive invocation "May Holy Mary and all the Saints intercede for us . . ." Even so, Pusey had to defend the sisters' use of a bowdlerised Roman Catholic Breviary in a letter to Alexander Beresford Hope: "It would not be true to say that it is the Breviary translated which they use. It is no other sense the Breviary than our English Prayer-Book is the Breviary" (Liddon 1893–4, III: 25). Pusey's reference to the "English Prayer-Book" is also illuminating, underlining as it does the fusion of church and state in the British psyche and the centrality of the *Book of Common Prayer* in British life.

Text and music

So the communities evolved, and each of them grew their own daily rituals based either on (Catholic) Latin texts or on Prayer Book words with additional English translations from the Latin. But a monastic liturgy and Office is very rarely a spoken ritual, so we must now add a musical twist to the fabric. Those 'Anglo-Papalists' who had chosen the Latin rite tended to consult their Catholic brothers and sisters and trained themselves to sing the chants to which the Latin words were wedded. The 'Prayer Book' communities, however, had no adequate musical resources and simply said the Office, but they soon found that singing was much more effective and supplemented the *Book of Common Prayer* by musical elements in manuscript. Back in the sixteenth century, Cranmer, realizing the need for musical settings, had commissioned Merbecke to set Prayer Book texts, and these were published as the *Boke of Common Praier Noted*; sadly this resource was no longer available, so the communities had to start from first principles. While there were some attempts to set the Prayer Book texts to classical melodies of the day (Beethoven, Mendelssohn, etc.) this genre proved unsatisfactory, and it was quickly realized that by far the best medium was the chant.

The chant had many advantages: not only was it tailor-made for the purpose, but it also provided a strong link with the long – and continuous – monastic tradition. But the only source of chant, to which the original texts were wedded, was the Roman rite, the music and ritual which had been used by Roman Catholic monastics since time immemorial. However, in terms of national identity, it was known that the Roman Rite itself had of course been used during England's Catholic past, so on one level there was a precedent for using the music. A second level of precedent also existed, because the rubrics of the *Book of Common Prayer* owed much to the rubrics of one particular local variant, or Use, of the Roman Rite. This was the Use of Salisbury, also known as Sarum Use, which was the predominant Use in pre-Reformation England: in 1923 Francis Eeles was to describe Sarum as the 'English Use', stating that whenever a manuscript is to be found describing itself as '*secundum usum Anglie*' (according to the English use) it will turn out to be of Sarum Use (Eeles 1923: 67).

English words to English music

In 1896 the Plainsong and Mediaeval Music Society produced a collection of chants, of which a revised (1937) edition became popular. For this, the music was mainly adapted from Sarum sources, ingeniously arranged "with considerable skill" by H. Briggs, Walter Howard Frere CR, and G.H. Palmer, to the words of the *Book of Common Prayer* (Allan et al. 1991: ix).

In this case, the texts were not Latin-English translations but the (albeit Jacobean) vernacular English words of the Prayer Book fitted to an existing chant repertory based on the 'English Use', by means of *contrafactum.* This appropriation of the words of the English Prayer Book, and the transformation of a textual medium into a sung medium, whether through translation and/or *contrafactum*, gave many communities in the ARC a sung Office, and the communities had a particular enthusiasm for something they saw as particularly English. As Father Peter Allan CR has said, "it is not too strong to say that there was not much sense of there being any choice", the liturgical material of the Use being peculiar to England and therefore appropriate.[7]

So the new monastic communities, each acting independently, started the long process of creating and adapting the music of Sarum to fit the English translations – or indeed vice versa. But the very act of Latin-English translation, and the difficulties of setting a translated text to an existing chant, meant that results differed widely: this can be illustrated by comparing just one example from the repertories of Community of St Mary the Virgin (CSMV) at Wantage in England and the Community of St Mary (CSM) in New York, USA.[8]

Each of these community repertories was based on the prodigious efforts of two clergyman-scholars, the Reverend Dr G.H. Palmer (for CSMV) and the Reverend Canon Charles Winfred Douglas (for CSM).

George Herbert Palmer (1846–1926) founded the chant tradition of both the Community of St Mary the Virgin (Wantage) and the Society of St John the Evangelist (Cowley), working from Sarum sources. His publications, mostly printed by the CSMV sisters on their own St Mary's Press, included *The Sarum Psalter: The Psalms and Canticles at Mattins and Evensong Pointed to the Eight Gregorian Tones from the Sarum Tonale* (1894), *The Offices or Introits for Sundays and Festivals with the Musical Notation from the Sarum Gradale* (1908), and *The Diurnal after the Use of the Illustrious Church of Salisbury* (1930). The new Episcopal communities in the USA benefitted from the work of American clergyman the Reverend Charles Winfred Douglas (1867–1944), who created English translations by taking "a little of everything"[9] but always working from the best primary sources available; he used both Sarum and Roman sources and was deeply influenced by the work of the Catholic monks at Solesmes. Foremost among Douglas's sources were the *Antiphonale Sarisburiense*, primarily a facsimile edition of a thirteenth-century manuscript edited by the Anglican monk-musicologist Dom Walter Howard Frere, and Volume 9 of the Solesmes *Paleographie Musicale*, a facsimile of the manuscript Lucca 601. Douglas's legacy is therefore drawn from disparate sources, rather than originating in a bipolar fashion from either Sarum or the Roman Rite.

Figures 6.1 and 6.2 show two versions of the introit 'Vocem jucunditatis'. The first (Figure 6.1) was written for CSMV (Wantage) by Dr George Herbert Palmer and was printed by St Mary's Press, Wantage; Palmer worked on the premise that the music came first and thus the music drove the adaptation of the language. Douglas's adaptation (Figure 6.2) is a version written for CSM (New York): it remains unpublished and is provided here in a manuscript copy, reproduced by permission of the Community of St Mary (East). Douglas maintained that "the sense and meaning of words must drive the music." A comparison of these two adaptations is illuminating: Palmer's version, although remaining in the Wantage repertory,

Figure 6.1 Palmer – Introit *Vocem jucunditatis*

Source: Copyright © Community of St Mary the Virgin. Used by permission.

Figure 6.2 Douglas – Introit *Vocem jucunditatis*

Source: Copyright © Community of St Mary. Used by permission.

challenges members of other communities, who have admitted they find it "problematic" and "difficult to sing",[10] whereas Douglas's aim was to provide a "simple, robust liturgy" for small communities in a form that suited the "natural English language inflection and cadence" for American English speakers.[11]

While both versions feature several identically set phrases, such as 'declare ye this, and let it be heard', the differing approaches of Palmer and Douglas have produced very different results in several places. For example, in the opening phrase 'with a voice of singing', Palmer's somewhat unfelicitous treatment emphasizes the words 'a' and 'of', while Douglas places the same *torculi* on the more significant 'voice' and 'singing'. Further on, in the phrase 'even unto the ends of the earth', Palmer's melismatic treatment of 'ends' creates problems with the short vowel sound, which Douglas avoids through the omission of several of the pitches. Douglas's version of 'Vocem jucunditatis' has therefore been pronounced melodic and eminently singable, demonstrating his premise that "the praise of God [should take] precedence over the virtuosity of the plainsong".[12]

Although the communities founded in the nineteenth century favoured the Sarum Use, and more and more useful Sarum-based chant editions appeared over the ensuing years, in the twentieth century a new enthusiasm for adopting Roman models became prevalent. One of the communities that chose to follow Rome in this way was the Society of the Precious Blood (est. 1905) at Burnham Abbey. They used a seven-fold Office of the Roman Rite, in Latin, with "the attitude to the Office [being] that it should be 'all-or-nothing' and great efforts [being] made to be 'pukka';"[13] in this context being 'pukka' meant using the *Liber Usualis*, the chant edition produced by the monks of the great French Catholic abbey of St Pierre at Solesmes and approved by the Vatican for official use.

Vatican II

In this way, Anglican communities continued to be founded, some using a Latin ritual and others firmly in English. But then, in the 1960s, everything changed with the Second Vatican Council, commonly known as Vatican II. Vatican II led to the Catholic Church effectively abandoning Latin in favour of a vernacular Office and liturgy, although this was neither intended nor anticipated by the Council. It affected many Anglican communities deeply. Those using the Latin Roman Rite, although not bound by the edicts of the Catholic Church, all chose to follow suit. For example, SPB (Burnham) continued to use the Roman Rite but created their own English translations, "picking and choosing the bits [they] wanted, in the Anglican way".[14] But translating into English meant that the chants needed to be extensively revised to fit the phrasing and accents of the vernacular, as can be seen in the following example.

Comparison of this adaptation (Figure 6.3) with the contour of the original chant in the *Liber usualis*[15] (Figure 6.4) shows the type of problems involved.[16] While the melodic contour is preserved as much as possible, comprehensive changes in accentuation and natural rhythm have necessitated alterations such as that in the first line (Figure 6.5).[17]

In this case, the problems raised by the need to preserve the sense of the opening phrase have led to the first few notes of the chant being completely omitted. The remainder of the first textual phrase now works well with a simplified version of the second musical phrase, albeit without the breath which divides the original. In the second line, problems are caused by differences in grammatical construction: in Latin the dative *discipulis* is used and the verb placed at the end of the sentence, whereas the English words require re-ordering as well as differing in word length and accentuation:[Figure 6.6] In each of the subsequent phrases, the English uses far fewer syllables than the Latin, such as the seventeen syllables of the final Latin phrase '*qui crediderit, et baptizatus fuerit, salvus erit*' set to thirty-six notes, compared to the eleven syllables of the vernacular phrase 'those who believe and are baptised shall be saved'. Such truncation of textual phrase length has resulted in the eleven syllables being set to only fourteen notes, providing a mainly

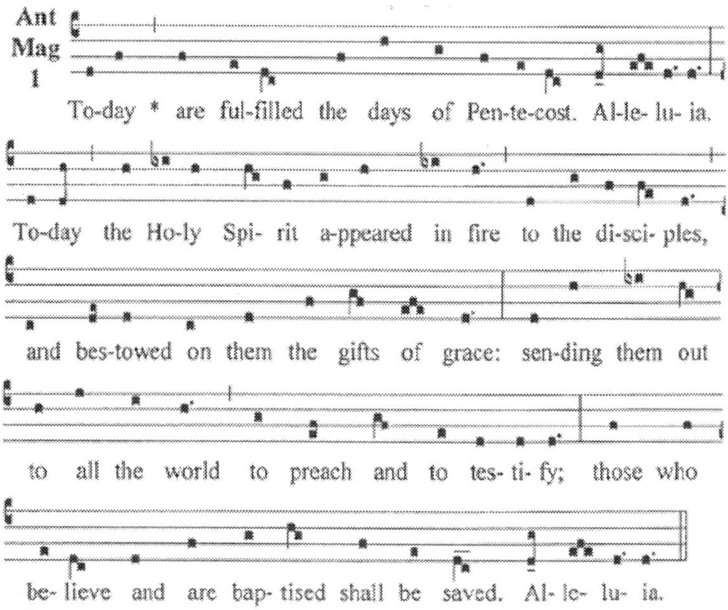

Figure 6.3 Magnificat Antiphon for Second Vespers on Whit Monday

Source: Copyright © Society of the Precious Blood. Used by permission.

Figure 6.4 Melodic contour of Magnificat Antiphon in Latin.

Source: Copyright © Amanda J. Haste.

Today* are fulfilled the days of Pentecost

˘ / ˘ ˘ / ˘ / ˘ / ˘ ˘

Hodie* completi sunt dies Pentecostes

˘ /˘ ˘ / ˘ ˘ /˘ ˘ ˘ / ˘

Figure 6.5 Natural stress features in opening of SPB Magnificat antiphon in English and Latin.

Source: Copyright © Amanda J. Haste.

Today	the Holy Spirit	in fire	to the disciples	appeared
Hodie	Spiritus Sanctus	in igne	discipulis	apparuit

Figure 6.6 Differences in word order between Latin source text and English translation.

Source: Copyright © Amanda J. Haste.

syllabic reading rather than the melismatic treatment which the preservation of pitches would have produced.

Given that each community's repertory may contain many hundreds of chants, and given that they were all using the same texts to a large degree, one might consider it simpler to adopt existing English texts which had already been translated for use in other communities, rather than undertake the enormous task of translation and adaptation all over again, rather like reinventing the wheel. But the fact is that each community's musical repertoire, developed over many years and used daily, becomes so integral to the community's life that it becomes a "collective representation" of the community (Asad 1993: 74–5) and speaks of the "inherited, singular springs of their identity" (Steiner 1993: 243). As a result of the laborious work of translation and chant adaptation, by the 1970s all Anglican communities were praying the Office in English. But as the language of Roman Catholic secular and monastic life was now also English, where did this leave the Latin language?

Latin as signifier

Latin is one of the languages closely associated with sacred texts and liturgies, which also include Sanskrit, Greek, and Hebrew, and it is through these sacred languages and written scripts that the "great sacral cultures" have historically "incorporated conceptions of immense communities" (Anderson 2006: 12). Many of these sacral languages (such as Latin) have traditionally become vehicular languages, functioning only as religious tools and symbols, but others have also begun to function as communicative tools: forms of Hebrew, for instance, are now developing into a communicative language in daily use, simultaneously functioning as signifiers of nationhood (as evidenced by the extensive literature on Jewish identity and national identity) and thus retaining – or rediscovering – a large element of communicative functionality.

Latin, however, is a 'dead' language which has not functioned as a communicative tool in general use for several centuries, and its use has long been confined to an elite of churchmen, classical scholars, lawyers, and others to whom it has been a professional tool. This elitism has raised its status to that of a prestigious language which, as a dead language, rarely tolerates the corruption suffered by vernacular languages as they evolve through daily – and common – communicative use. According to Anderson, if a language is to function as a sacral language, "the deader the written language – the farther it [is] from speech – the better" (2006: 13). This is not least because such a dead language is by definition static, reliable, incorruptible, and thus prestigious, the prestige being "located in a language because it is the preferred vehicle of religious truths" (Bellos 2011: 172).

While Latin has lost none of its prestige – and as fewer and fewer people have any education in the classics, it has perhaps become even more prestigious – once abandoned in Catholic worship, the language lost its most important function in the modern world: that of a "vehicle of cultural expression" (Armstrong 2005: 21), in this case a language infused with Roman Catholic religious culture. The loss of this significative potency is vividly illustrated by the fact that members of the ecumenical Taizé community were able to choose Latin as their ritual language when they commissioned a new musico-textual repertory, on the grounds that as it was no longer used by any one nation it belonged to no one at all. The Taizé Community is an international ecumenical monastic order in France (est. 1940) and is made up of more than a hundred brothers, from Protestant and Catholic traditions, from more than thirty countries across the world. Taizé attracts many visitors, and services may be attended by up to forty nationalities, so in order to overcome the problem of exclusivity in an international community, the monks commissioned the composer Jacques Berthier to produce an entirely new repertory for them and their visitors. Over the next twenty-five years, Berthier produced a corpus of 232 chants to Latin texts: although they have all since been translated into many living world languages, it is the Latin versions that are reproduced on Taizé recordings and playlists and which have remained the *urtext* and defining *lingua franca* of Taizé music around the world.

The interesting point here is that Berthier was himself Catholic, and this poses the question of the actual status of Latin as a functional language within the Catholic Church in the late twentieth and early twenty-first centuries. Ulrich Matthias gave an elegant synthesis of the position when he recalled that in 1999 Cardinal Ratzinger "called on the bishops to 'recover the Latin Mass'" because "[in] his view the 'wild creativity' that followed Vatican II 'destroyed the mystery of the Sacred'";[18] and also that in 2002 Pope John Paul II "emphasized that Latin remains the official language of the Catholic Church", pointing out that the use of Latin "is an indispensable condition for a proper relationship between modernity and antiquity, for dialogue among different cultures, and for reaffirming the identity of the Catholic priesthood."[19]

But, as Matthias said, "[t]hings have changed" and "the Latin language has shown itself to be too difficult to serve effectively the unity of the Church" (Matthias 2002: 41): even the pope's latinist, Abbot Carlo Egger, is reported to have said: "Latin now stands very little chance of survival in the Catholic Church. The simple truth is that many, too many, bishops no longer know how to speak it".[20] But is Latin well and truly dead to the Catholic Church? Certainly very few can now read or write it, but as a symbol of Catholicism it is still showing a flicker of life: in 2001 the German news agency DPA reported that, at a meeting of 155 cardinals in Rome in May of that year, Cardinal Jorge Maria Mejia from Argentina[21] "began his address in Latin, thereby causing confusion especially among the interpreters. When he continued his speech in Spanish, the audience applauded. 'Don't be frightened', he said, 'I do this only so that Latin may have at least a symbolic presence in this hall'."[22]

(Re)Drawing the boundaries

Such statements reveal an understanding that Latin has now lost its functional role as a religious signifier, becoming merely a symbol rather than a useful communicative tool. As both Catholic and Anglican communities now use English, this means that, on the linguistic surface, there is nothing to choose between Catholic and Anglican monastic communities. Their respective rites have of course diverged to reflect the dogmatic details on which they differ, and there are issues which divide them still, such as women's ordination, priestly celibacy, and attitudes to divorce, although the words they all use to express the basic tenets of Christian theology stem from the same roots and are all in the same language. But does this use of the vernacular mean that their shared Catholic heritage can no longer be heard in a physical sense? After all, the pre–Vatican II Latin-English divide between the former 'Anglo-Papalists' and the 'Prayer Book' communities in the Anglican denomination served as a daily reminder of which voice they heard more clearly. Have the communities now lost the "natural semantics of remembrance" (Steiner 1993: 494)? The link between language and their religious identity is inextricably entwined with their sense of national identity, a phenomenon still in evidence in the 1990s: when a group of monastic musicians and secular scholars compiled *An English Kyriale* (1991), they "usually (but not exclusively) followed the Salisbury musical readings", because "[i]t has seemed to us valuable to preserve an 'English' character for the collection" (Allan et al. 1991: x).[23]

As of now, Anglican and Catholic religious communities have been singing and praying in the same vehicular language of English for more than fifty years, and Latin has been effectively consigned to the archives. Whether or not they have lost their "language-identity linkage" (Edwards 2009: 54), their shared history and common liturgical practices are now accepted – *lex orandi, lex credendi* ('As we pray, so we believe') – and in the Vatican decree on ecumenicisism *Unitatis redintegratio* it was emphasized that "[c]ommon prayer forms part of this inheritance still shared by Anglicans and Roman Catholics despite centuries of division" (French Anglo-Roman Catholic Joint Committee 2015: 8). Even so, some Anglican religious find the voice of the Catholic Church still calling to them, and there is a steady trickle of religious who 'go over' to Rome, finding the call of their Catholic heritage stronger than their sense of Anglican identity. The latest – and most notable – of these secessions involved one of the oldest and most prominent sisterhoods in the ARC when, after a long period of discernment, a significant number of the sisters of CSMV (Wantage) – along with one sister formerly of the Society of St Margaret at Walsingham – joined the Personal Ordinariate of Rome, becoming a new Roman Catholic community known as the Sisters of the BVM. The Catholic press revelled in this 'conversion':

> the fact that the Community is, in its Anglican manifestation, now in its closing years (for it can be said with some certainty that the Anglican community will have come to an end with the death of its last sister, and that any new vocations will be to the Catholic community now crossing the Tiber) mirrors exactly the historical situation of that part of the Anglo-Catholic movement which remains within the Church of England. This will become progressively more enfeebled as the Church of England itself becomes more and more secularised: and it is now, surely, time for those within it still trying to live their lives within the Catholic tradition to come home to the place that has been prepared for them.
>
> (Oddie 2012: n.p.)

The shock waves generated by the secession to Rome of this historic community are in no small part a reflection of their status within the ARC and the fact that their huge, complex repertory based on the Sarum rite has become synonymous with an 'English' – meaning Anglican – rite. In joining the Catholic Church, the sisters have abandoned many of the tenets of twenty-first-century Anglicanism: for instance, one ordained sister has revoked her priesthood as being incompatible with Roman Catholic doctrine. So will they be abandoning their musico-textual repertory in favour of a Roman Catholic one? It seems not. An article in *Anglican Patrimony* in December 2012 says: "These twelve sisters will [. . .] continue in their work of prayer and contemplation, while retaining certain of their Anglican traditions and practices. Foremost amongst these is the tradition of English plainchant for which these sisters are well known" (*Anglican Patrimony* n.p.). The 150-year journey of the ARC to date provides ample evidence of the translational processes involved in identification, which is, in the words of Stuart Hall (1996: 2–3):

> a process of articulation, a suturing, an over-determination not a subsumption. There is always 'too much' or 'too little' – an over-determination or a lack, but never a proper fit, a totality. Like all signifying practices, it is always subject to the play of *différance*.

Conclusions

As we have seen, the issues facing Anglican religious communities can be seen as a web of enmeshed identities, in which the triple strands of Englishness, Anglicanism, and Catholicism are entangled. The autonomy and freedom of choice enjoyed by the ARC has allowed them to express their *différance* by choosing and adapting resources which speak to them as Anglicans, and as English men and women; they have been able to cross and re-cross boundaries, seeking legitimacy by drawing on sources perceived to be historically valid, but this process has in turn presented them with the practical problems of translation and the considerable challenges involved in setting English texts (whether English translations from Latin or the vernacular English of the Prayer Book) to chants intended to accommodate the rhythm and cadences of Latin.

The communities themselves are dynamic entities, evolving and changing with the individuals who constitute them, and their repeated efforts to align themselves to their Church, and all those who have lived the English religious life over the centuries, whether Catholic or Anglican, continue to evolve and change in response to changing social conditions and religious dogma. One thing is sure: while the Latin language has provided a singular link with their Catholic heritage, that linguistic link has now been well and truly broken, and all their future responses will be expressed in the vernacular . . . as will those of their English Catholic brothers and sisters. Where this will lead them all remains to be seen.

Notes

1 *Convents in Exile 1600–1800*, edited by Caroline Bowden (Pickering and Chatto, 2012–13) provides an excellent insight into spiritual and practical life in the post-Reformation English Catholic monastic diaspora.
2 In 2013 there were some two thousand monks, nuns, and friars in the Anglican Communion. Source: Anglican Religious Communities, *Anglican Religious Life 2014–15.*
3 'The religious' is a generic term encompassing all those living a vowed monastic life, such as monks, nuns, and friars.
4 Although 'monastery' is a term which can be applied to monastic communities of both genders, here I am using 'convent' and 'monastery' to denote female and male communities, respectively.
5 The Sacramental Test Bill of 1828 repealed the Test and Corporation Acts and restored the rights of Catholics to hold public office.
6 These included the *Day Hours of the Church with Gregorian Tones, translated from the Sarum Day Hours* by the Reverend Albany Christie (1844), the *Divine Office of the Holy and Undivided Trinity* (1872) and the *Day Hours and Other Offices as Used by the Sisters of All Saints* (1875). In 1852, a book was published in London entitled *The Psalter, or Seven Ordinary Hours of Prayer, according to the Use of the Illustrious and Excellent Church of Sarum*, which contained many of the old Offices, and plainsong melodies for Chambers' translations of the Office Hymns. The chant was crudely and experimentally adapted to the English text, but the work was a brave and pioneer effort in the wider efforts to revive the use of plainsong in the Anglican Church.
7 Letter to the author, 5 August 1997.
8 Although working to provide an Office for an American convent, Douglas's work has been included here because he showed the same concern for using 'English' sources as did Palmer in England.
9 Mother Miriam CSM, email to the author, 19 March 2009.
10 Comments to the author at the Anglican Liturgical Musicians conference, May 2007.
11 Mother Miriam CSM, email to the author, 19 March 2009.
12 Mother Miriam CSM, email to the author, 19 March 2009.
13 Sr Margaret Mary SPB, personal interview with the author, 21 January 2006.
14 Sr Margaret Mary SPB, personal interview, 21 January 2006.
15 This antiphon appears in square notation in the 1950 edition of the *Liber Usualis* (Desclée de Brouwer, Paris): 886.
16 In this realization, a dot after a note denotes a longer note, while a dash above or below a note denotes an accented syllable.
17 This setting treats 'Hodie' with unexpected accentuation: as Sr Margaret Mary SPB says: "When one says *hodie* the accent is on *'Hod-'* but in each of

the antiphons where it is sung the chant puts the accent on the '*di*'." [Email to the author, 23 February 2009.]

18 "Die Messfeier in lateinischer Sprache wiederentdecken," *Die Welt*, 22 December 1999, cited in Matthias 40.

19 "Pope Pushes for Wider Use of Latin", *EWTN News Brief*, 22 February 2002, cited in Matthias, 41.

20 James Meek, "Latin is all Greek to us, Catholic bishops admit," *The Guardian*, 16 October 1999, 5, cited in Matthias 41.

21 Cardinal Mejia served on the Board of World Religious Leaders for the interfaith dialogue-based organization the Elijah Interfaith Institute from 2013 until his death in 2014.

22 "Leb wohl, Latein – Sprachprobleme beim Konsistorium in Rom," *DPA*, 24 May 2001, cited in Matthias, 41.

23 *An English Kyriale* follows in the path of the PMMS *Ordinary of the Mass* (1896, revised edition 1937).

References

Allan, Peter C.R., Berry, Mary, Hiley, David, Pamela, C.S.J.B., and Warrell, Ernest, editors. *An English Kyriale: Music for the Eucharist*. London: Community of the Resurrection, Harper Collins, 1991.

Anderson, Benedict. *Imagined Communities: Reflections on the Origin and Spread of Nationalism*. London and New York: Verso, 2006.

Anglican Patrimony. Blog. Posted by Fr. Bartus. 3.47 p.m., 12 December 2012. http://anglicanpatrimony.blogspot.co.uk/2012/12/anglican-nuns-convert.html

Anglican Religious Communities. *Anglican Religious Life 2014–15: A Yearbook of Religious Orders and Communities in the Anglican Communion, and Tertiaries, Oblates, Associates and Companions*. Norwich: Canterbury Press, 2013.

Armstrong, Nigel. *Translation, Linguistics, Culture*. Topics in Translation. Edited by Susan Bassnett and Edwin Gentzler. Clevedon, Buffalo, and Toronto: Multilingual Matters, 2005.

Asad, Talal. *Genealogies of Religion: Discipline and Reasons of Power in Christianity and Islam*. Baltimore, MD: Johns Hopkins University Press, 1993.

Bellos, David. *Is That a Fish in Your Ear? The Amazing Adventure of Translation*. London: Penguin, 2011.

Benedictines of Solesmes; Society of St John the Evangelist, eds. *Liber Usualis with Introduction and Rubrics in English*. Tournai: Desclée, 1950.

Bowden, Caroline, general editor. *English Convents in Exile, 1600–1800*. Volumes 2 and 3. London: Pickering & Chatto, 2012–3.

Edwards, John. *Language and Identity*. Key Topics in Sociolinguistics. Edited by Rajend Mesthrie. Cambridge: Cambridge University Press, 2009.

Eeles, Francis C. *Prayer Book Revision and Christian Reunion: Some Essentials Discovered in the Light of Liturgical Facts and Principles*. Cambridge: Cambridge University Press, 1923.

French Anglican-Roman Catholic Joint Committee. *For a Common Prayer between Anglicans and Roman Catholics: "O Lord, Open Our Lips."* [Documents épiscopat]. Paris: French Catholic Bishops' Conference, 2015. https://iarccum.org/archive/France/2014_french-arc_for-a-common-prayer_en.pdf

Hall, Stuart, and Du Gay, Paul. *Questions of Cultural Identity.* London: Sage, 1996.

Hobsbawm, Eric, and Ranger, Terence. *The Invention of Tradition.* 1983. Cambridge: Cambridge University Press, 2012.

Joseph, John E. *Language and Identity: National, Ethnic, Religious.* Basingstoke and New York: Palgrave Macmillan, 2004.

Liddon, H.P. *Life of Edward Bouverie Pusey.* 4 volumes. London: Longmans, 1893–4.

Matthias, Ulrich. *Esperanto – The New Latin for the Church and for Ecumenism.* Translated from Esperanto by Mike Leon and Maire Mullarney. Preface by Dr György Jakubinyi, Archbishop of Alba Iulia, Romania. Antwerp: Flandra Esperanto-Ligo, 2002.

Oddie, William. "The Wantage Community Was One of the First Anglican Religious Orders; The Conversion of Its Core Members Heralds the End of Anglo-Catholicism". *Catholic Herald.* Blog. Posted on 18 December 2012. www.catholicherald.co.uk/commentandblogs/2012/12/18/the-wantage-community-was-one-of-the-first-anglican-religious-orders-the-conversion-of-its-core-members-heralds-the-end-of-anglo-catholicism

Philogene, Gina. "Social Representations of Alterity in the United States". In *Social Representations and Identity: Content, Process, and Power*, edited by Gail Moloney and Iain Walker, 31–42. New York: Palgrave Macmillan, 2007.

Steiner, George. *After Babel: Aspects of Language and Translation.* 2nd edition. Oxford: Oxford University Press, 1993.

7

BIBLICAL SEMIOTICS AND THE POLITICS OF SURVIVAL

Gilad Elbom

Introduction: warning signs

In the early 1980s, Thomas Sebeok was asked by the United States Department of Energy to propose a fail-safe semiotic solution to prevent human interference with nuclear waste repositories. The purpose was to assure that future generations would stay away from geological sites where lethal substances are stored. Since these substances are expected to remain radioactive for more than ten thousand years, the main concern was to design a warning system that would be effective long after established methods of communication, including currently known languages, ceased to be understood. Considering the fact that no civilization in history has survived for ten thousand years, the assumption was that with the passage of time, knowledge about the original content of a nuclear waste repository would inevitably be lost. In such a case, a periodic ritual based on legend and superstition rather than accurate information might be the most effective way, Sebeok suggested, to make sure that hazardous areas remain off limits (1986: 168). An elite group of experts – an atomic priesthood, so to speak – would be exclusively acquainted with the truth behind the myth, the danger behind the taboo, the possibility of radiation, and its implications. The rest of the population – the uninitiated – would simply follow a tradition.

In semiotic terms, a radioactive site whose deadly qualities cannot be named, specified, or recalled is a signified without a signifier. In other words, the initial warning sign – DANGER: NUCLEAR WASTE – would be replaced with a series of substitutes, evolving and changing through time, until its original meaning would be no longer comprehensible. The only thing that would remain, at least among the uninitiated, is the notion of some dreadful site, object, or entity that should be forever treated with awe and caution.

One of the religious equivalents of such an entity – a concept without representation – is the Hebrew deity. Guarded by an elite priesthood, the very name of God has been erased, leaving multitudes of practitioners with nothing but a nebulous notion of the object they worship: a referent without a sign. Other vehicles that could imaginably signify this divine concept have been outlawed: statues, paintings, figures, masks, and all other types of iconic representation. Over the course of more than three thousand years, the Jewish priesthood has disappeared, and with it the vocalization of the name of the deity. The deity itself, however, survived, as did Jewish culture. Some would argue that what has facilitated this survival – the deliberate cultivation of vagueness and abstraction, accompanied by daily rituals based on repetition rather than information – is precisely the kind of communication envisioned by Sebeok: the successful perpetuation of an important myth despite an obvious lack of accurate data about its exact appellation, nature, shape, dimensions, or powers. Others, however, might be more inclined to suggest that it was active resistance to the idea of the priesthood that has played a more effective role in the survival of Jewish civilization.

Garfield (1994: 15), in her critical response to the concept of an atomic priesthood, protests against Sebeok's "pervasive condescension towards the populace, kept in ignorance about radioactive burial sites millennium after millennium," claiming that such an attitude is based on "secrecy, manipulation, and deceit." Houser (2007: 11), impressed as he is with the ingenuity of the idea, expresses similar reservations. "I too," he says, "am less than delighted with the thought that the best way to transmit important knowledge from generation to generation, for at least ten thousand years, is occultly through the vague semiotic content of religious practices."

The interplay of a cloak-and-dagger clergy in charge of national secrets and critical voices that call for transparency and demystification is one of the central aspects of the textual tradition of biblical and rabbinical literature. In its capacity as a complex mechanism meant to guarantee the survival of the nation, this textual tradition proposes its own semiotic doctrine. With continuity as its primary characteristic and objective, this doctrine of signs is very similar to later formulations of semiotic models. "If we ask what it is that semiotic studies investigate," says Deely (1990: 22), "the answer is, in a word, action. The action of signs." Sign action, also known as semiosis, is "constantly renewed and presupposes continuity" (Proskurin 2010: 21). This continuity relies on the interchangeable roles of the three elements comprising the sign: the representative element, also known as signifier, sign-vehicle, or representamen; the signified element, also known as object, referent, or significate; and the point of interpretation, also known as interpretant, which occurs, according to Peirce (CP 8.177), when an interpreting mind, actual or potential, observes the relation between the two other elements. "The interpretant is the third component of a sign, and is itself a third, in that it consists in a meaning which emanates from the connection between sign and object" (West 2012: 313–14). Following the triadic model suggested by Peirce, Deely (2010: 96) offers an elegant summary of sign action:

> The representative element within this triadic structure, which we loosely call a "sign," "in itself" is not a sign at all, but one of the three elements necessary to the being of a sign, one of the three legs on which the sign walks in working its way through the world, and, indeed, the "foremost" leg, insofar as it is the leg which takes the direct representative step carrying a semiosis. What is sign-vehicle one time can be significate another time; and what is interpretant one time can be sign-vehicle the next time; and so on, in an unending spiral—an "unlimited semiosis."

The principle of infinite sign action is crucial to the history of the Hebrew priesthood and the political elements that challenge it, especially when key figures in biblical and rabbinical literature are inspected as elements that transform, evolve, and turn into new signs: Moses and Aaron, the founders of the Hebrew priesthood; critical voices that strive to subvert the power of myth, especially in Exodus and Numbers; God, who seems to be trapped between political rivals who wrestle for his validation; Jesus, who leads a failed campaign against Temple authorities; and Yohanan ben Zakkai, who puts an end to the political reign of the priests. In other words, it is the tension between mutually reliant constituents, not the advantage of one when measured against the other, that contributes to the development of early Hebrew semiotics and secures the stability of Jewish civilization. Ultimately, continuity is established not through the "legend-and-ritual" of a "commission of knowledgeable experts" (Sebeok 1986: 168) but as a product of the dynamic opposition between those who support it and those who defy it.

Moses: the great encoder

The distinction between a small layer of privileged experts and a large population of inadequately informed and strictly governed subjects is one of the central conflicts in the story of Moses. As if against his will, Moses is initially called to save the people from their Egyptian oppressors. The problem, of course, is the ostensibly odd decision to disclose what otherwise looks like classified information: the Egyptians appear to be helpless puppets manipulated by a higher power operating behind the scenes. The consistent lack of transparency may be seen as a striking precursor to the intrinsic vagueness that Sebeok identifies as a vital component of mythmaking.

Among the important anti-opposition laws that Moses introduces are those concerning the distribution of natural and human resources, the regulation of sexual and religious practices, and the ban on private magic (Exodus 22:18; Leviticus 19:31; 20:27; Deuteronomy 18:10–2). It would be interesting to examine the ways in which Moses and the Levite priesthood guarantee their superiority. The first plague that Moses inflicts on Egypt – the poisoning of the Nile – is preceded by a live demonstration of magic and so-called black magic. And this is the moment that marks the entry of Aaron, soon-to-be first official high priest, into the political scene. Moses and Aaron present an extravagant production of authorized magic: poison in the name of God. Contrasted with the illegitimate poison of the Pharaoh's sorcerers, the magic of Moses and his executive brother is not a public health hazard but a benign force that serves and protects.

A subtle emphasis on critical perspectives can be detected when an inexplicable plague claims the lives of thousands of people (Number 25). Moses, who appears to be concerned with the purity of the community, is quick to identify the disease as collective punishment for sexual relations between a Hebrew man and a Midianite woman. Phineas, a loyal grandnephew, slaughters the delinquent couple, the plague stops, and eternal priesthood is granted to the descendants of Aaron. Exposing the hypocritical nature of the leadership, the biblical narrator insists on mentioning, and more than once, that Moses himself is married to a Midianite woman (Exodus 2:15–22; 18:1–12; Numbers 10:29–34; Judges 4:11).

Perhaps it is no coincidence that subversive voices in the narrative – voices that suggest alternatives to the iron fist with which Moses rules the nation, voices that call for greater degrees of dialogue and democratization – are minor, subtle, or altogether absent. Semiotically speaking, such voices belong to a different doctrine of signs: dynamic, interpretive, dialogic. Often associated with Peirce or Bakhtin, this type of doctrine, sometimes called *interpretation semiotics*, is based on "constant regeneration of interpretation" (Proskurin 2010: 21). In contrast, the semiotic model offered by Saussure, termed by its critics *decodification semiotics* or *equal exchange semiotics*, is based on binary oppositions. The shortcomings of *decodification semiotics* are enumerated by Petrilli (2015: 61):

> According to this model, the sign is: 1) at the service of meaning pre-established outside communication and interpretation

> processes; 2) considered as a pre-constituted and passive instrument in the hands of a subject who is also given and pre-established antecedently to semiosic and communicative processes, therefore capable of controlling and dominating signs and sign processes at will; 3) can be decoded on the basis of a pre-existing code shared by partners in the communicative process.

In this framework, signification is reduced to a purely mechanical operation: the recipient of a transmission uncovers the content that hides behind an expression and reveals its intended message. Devoid of the ability to acquire new meanings, "the sign is dominated by the logic of perfect correspondence between that which is given and that which is received, that is, by the logic of equal exchange which currently regulates all social relations in today's dominant economic system" (Petrilli 2015: 59–60). In a sense, this is the model that Moses imposes on the people. The leader is a vehicle that signifies divine intentionality – the unchanging will of God – and communication is reduced to the binary oppositions of obedience and reward, defiance and punishment, life and death. It is only later, during the reign of Saul, the first king, that a different semiotic model is proposed.

A more dynamic model is practised, very briefly, by Zipporah, the Midianite wife of the leader-to-be. Realizing that circumcision is the mark of the covenant between the human and the divine, she performs a ceremony that appeases a bloodthirsty God and saves the life of her husband (Exodus 4:24–6). Stopping short of treating this enigmatic pericope in semiotic terms, Wildavsky (1984: 48) recognizes the triadic and interchangeable nature of sign relations: "Zipporah, in a tripartite ceremony, marries God and Moses to the people, God and the people to Moses, and the people and Moses to God." But how does she guess why God wants to kill Moses? How does she know that circumcision is the problem? Rabbinical literature – especially the tradition of midrash: the kind of textual expansion that acknowledges the elliptical and elusive nature of biblical stories, identifies the absence of important information, and transforms the original narrative into a memorable folktale (אגדה) – proposes numerous inventive explanations, among them the ideas that it was neither Moses nor his firstborn son, Gershom (Exodus 2:22), that God wanted to kill but Eliezer, the younger son (Exodus 18:4), and that it was not God himself but an angel of God. According to Rashi – Rabbi Shlomo Yitzchaki (1040–1105), whose extensive commentary on the Torah often borrows from the imaginative tradition of midrash – the angel of God turned into a snake and swallowed the boy from his head to his loins, then swallowed him again from his feet to the same spot in his body, indicating the problematic area.

It was then that Zipporah understood what she had to do. Aichele (2001: 141), who approaches the story from a semiotic perspective, suggests an intertextual explanation:

> Uncertain denotation is here produced by an insufficiency of the signifier. Nevertheless, Zipporah has understood what the book's reader cannot. Somehow she alone knows just what to do. These gaps in the narrative must be filled in by the reader, and Zipporah is a very good reader. It is as though she has already read the larger story in which she appears to be a minor character. It seems that she already knows the canon—at least the canon of Torah. Once again the text loops back on itself and presents an aesthetic message. Zipporah's story both invites and staves off the threat of semiotic chaos.

The text itself functions as a never-ending meaning-making game: a choose-your-own-adventure narrative. The task of the players, characters and readers alike, is to suggest creative interpretations to vague, incomplete, or disembodied signifiers. In other words, the task of the interpreter is to create context, not to adhere to prescribed notions of chronological or linear reading. In Semiotic terms, "the meaning of a text does not lie in the text itself but in the juxtaposition of texts (representamen and object), which itself can be justified only in terms of some further arrangement of texts (interpretant)" (Aichele 2001: 149).

Moses himself, as a highly self-conscious literary figure, appears to understand, like his wife, the power and control that come with intertextuality. Communicating his awareness of the fact that he is the most indispensable character in the narrative, he asks to be erased from the very book in which he appears (Exodus 32:32). These self-referential moments are later appropriated by characters, readers, and readers who themselves become important textual characters: the experimental exegetes of rabbinical literature. As a collective of commentators, they construct a more flexible semiotic model, one that shifts the focus from encoder to interpreter and grants everybody, not just the leader and his family, the freedom of arrangement and rearrangement.

Jesus: the new Zerubbabel

As a national symbol, the Temple in Jerusalem, envisioned by David and built by Solomon, becomes the quintessential representation, both physical and conceptual, of Jewish sovereignty. While the priests and Levites are the ones who operate it, the Davidic monarchs – first in actuality, then in the collective consciousness of the people – are its supreme proprietors. When the Hebrew nation has lost its independence and the Davidic dynasty is no longer in power, the yearning for a messiah – a new Davidic king – takes the form, among other things, of organized attacks on the priests, whom large portions of the population, especially during the Roman occupation, see as domestic collaborators with the foreign powers that have deprived the nation of its political, cultural, and religious autonomy. Supported by the people, Jesus tries to eliminate priestly corruption, perhaps naïvely, when he enters Jerusalem, launches a one-man campaign against Temple authorities, invades their headquarters, challenges their position of power, overturns the tables that signify the commercialization of worship, and declares that the Temple itself will be destroyed and replaced with a new one. Ultimately, a much more effective act of resistance is the Pharisaic idea that the Law – the text, the Bible – is not the exclusive property of the priests but a public database that the people are encouraged to study, debate, interpret, and appropriate. And that is exactly what Jesus advocates prior to his open confrontations with the priests. As political signifiers, his legal disputations with the Pharisees demonstrate the personal responsibility of every Jew to develop a close familiarity with the layered complexity of the text rather than rely on the type of rigid, unimaginative interpretation associated with the Sadducees.

In other words, Jesus of Nazareth operates as part of a decentralization movement. His public debates with the Pharisees, fellow activists in the same movement, are performative exercises that display and assert the right and ability of unordained citizens, especially in the Galilee, the desert, and other peripheral areas, to offer autonomous alternatives to the official centre of power in Jerusalem. Local synagogues and impromptu gatherings replace the formal setting of the Temple. The local rabbi feeds the masses rather than masses of pilgrims feeding the priests (Matthew 14:13–21; 15:32–8; Mark 6:30–44; 8:1–10; Luke 9:12–17; John 6:1–14). Complicated and expensive rituals are simplified, demystified, and reclaimed by the people. Responsibilities are shared, and ongoing conversations replace the final decisions of a small clerical circle.

The question of forgiveness is a prominent example. According to Jesus and the Pharisees, it is the responsibility of individuals to seek and approach, in an open and direct manner, those whom they have wronged, without the mediation of the religious establishment, and beg for personal forgiveness (Matthew 5:23–4; 6:12; 6:14–15; 18:15–17; 18:21–2; Mark 11:25–6; Luke 11:4; 17:3–4). The very idea is another assault, and a most successful one, on the hegemony of the priests. As such, it proposes a popular method to circumvent the obligatory guilt offering at the Temple in Jerusalem (Leviticus 6:1–7 [5:20–6 MT]). A clear case of taxation in disguise, this type of sacrifice offers the people an unreasonably expensive service that they are perfectly capable of performing themselves, especially when the victim of the offense is a fellow human being, not God. In political terms, Jesus and the Pharisees protest against the commodification and monopolization of official repentance procedures, as well as the unnecessarily mysterious nature of priestly work.

As political acts, the parables, and especially the refusal to provide authorized explanations or definitive interpretations, signify not only a rabbinical alternative to the thesis-based, persuasive argumentation of Greek rhetoric but also a more equal distribution of hermeneutical privileges among observant Jews. Questioning the axiomatic claim that only the priest can speak on behalf of God or communicate the meaning of Scripture, Jesus teaches his immediate audience to play an active role in the interpretation of their own textual and oral tradition. According to Jesus, no authority is immune from critical reassessment, not even Moses (Matthew 5:31–2; 19:3–9; Mark 10:2–12; Luke 16:18). His critique of Moses allows Jesus to position himself in dialogue with the richness and complexity of the biblical text, echoing and expanding the polyphony of the Hebrew Bible. The characterization of Temple authorities as irredeemable villains, especially alongside the portrayal of Jesus as the long-awaited anointed king, can be seen as an inevitable development of that textual tradition, most notably the visions of Jerusalem in Ezekiel (8–9) and Zechariah (3–4), as well as the chastising of the priests in Malachi (1:6–3:15) and the deuterocanonical Testament of Levi (14–18).

According to Ezekiel, the corruption of centralized rituals is orchestrated by the elders of the house of Israel (8:9–16): the ruling elite at the Temple. Vivid descriptions of idolatrous practices are followed by the promise of purification through death and destruction (9:4–10), then rebuilding and reunification. According to some visions, the future leader will be a king: a descendant of the house of David (37:15–28). According to others, a righteous priest: a descendant of the house of Zadok (40:45–6; 44:15–16; 48:11). The duality of king and priest reaches a more explicit level when Zechariah names the rival authorities: Joshua, the high priest, is wearing filthy clothes, standing in front of an angel of God, humiliated by Satan, the accuser (3:1–4). God forgives his sins (3:4) but entrusts Zerubbabel, a glorious descendant of the house of David, with the task of rebuilding the Temple (4:8–10). Escalating the case against the priests, Malachi and the Testament of Levi offer scathing bills of indictment that portray the Levites in charge of the Temple as crooked, conceited, idolatrous, avaricious, and sexually depraved.

The picture that this textual tradition paints is clear: Jerusalem – the dwelling place of the Hebrew deity (Joel 4:16–17; Ezra 1:2–3) – is soiled. And if this is the case, where does God reside? In a heavenly Jerusalem, the

Jerusalem above. The priests, who admit that they have been unfaithful custodians of the Temple, throw the building keys into heaven and surrender them to God (2 Baruch 10:18; *b.Taanit* 29a; *j.Shekalim* 26a–b). God reaches out his hand and takes the keys, along with all the other holy vessels of the tabernacle, and keeps them in his heavenly kingdom (2 Baruch 6:7–9). According to popular apocalyptic visions, this heavenly kingdom, the New Jerusalem, will descend from above and replace the occupied, corrupt, or destroyed one (2 Esdras 7:26; 10:38–55; Galatians 4:25–6; Revelation 3:12; 21). In theological and political terms, Jesus and the Pharisees, who blame the priests for the desecration of the Temple, call for the kind of emancipation that will replace not only the human with the divine, current history with the world to come, or imperialist occupation with Hebrew autonomy but also, and on a more immediate level, the reign of the Sadducees with divine-sanctioned self-reliance.

The call to those who are guilty of insulting their friends to abandon the offering at the altar and appease their victims directly, without the mitigation of clerical rituals or intervention of judicial authorities (Matthew 5:21–6), could be seen as a turning point in the formation of a political platform. No longer does Jesus preach indirectly against the official narrative of priestly hegemony. From this moment, he describes new procedures and practices that will eventually render the priests redundant. These new practices are, once again, legitimized with divine endorsement. Horizontal forgiveness – the kind that is shared among the people and ultimately circumvents the priests – will be followed by forgiveness from God. Vertical forgiveness – the kind that forces the sinner to appeal to the priest, who then appeals to God – will not produce forgiveness from above unless preceded by the vertical procedures described here (Matthew 6:12–15; Luke 11:4). In this context, the fact that Jesus heals a leper and commands him to go and make the customary offering to the priest (Matthew 8:4; Mark 1:44; Luke 5:14) might be self-contradictory. Nevertheless, if we accept the notion of an organized opposition to a political elite, it might be possible to read this command as ironic. The idea of offering thanks to the priest for a miracle performed by Jesus is laughable, and Jesus might be overtly sarcastic here.

The emphasis on active participation expresses a yearning for a more reciprocal model of sign relations. The priesthood no longer controls the text mystically, nor does the monarchy control it by force. No longer does the leadership send strict commands or unequivocal messages to passive hearers through consecrated, incontestable, eternal national symbols. As signifiers, Jerusalem, the Temple, and the text – the Torah – are appropriated by the people, reinterpreted, and replaced with more important components in the semiotic process: the interpreters themselves. This is exactly the process that Petrilli (2015: 63) describes when she explains the distinction between decodification semiotics and interpretation semiotics:

> The interpreter/interpretant responds to something and in so doing becomes a sign which in turn gives rise to another interpretive response, etc. From this perspective, the function of the interpretant sign is not limited to merely identifying the previous sign, but rather is taken to various levels of responsive understanding (or answering comprehension) which implies the existence of a concrete dialogic relationship among signs regulated by the principle of reciprocal otherness.

The replacement of the Torah with the Oral Law – the Mishnah, the Talmud, and rabbinical literature in general – may seem controversial, even heretical. After all, what could be more important than the Torah? In fact, this type of substitution is an inevitable development in the transition to interpretation semiotics. What matters more than the Torah is the study of the Torah. From a historical viewpoint, the Sadducees, who rejected the Oral Law, were bound to disappear. Those who recognized the centrality of the interpretant – the Pharisees, who later evolved into rabbinical and scholarly Judaism, both religious and secular – understood that the best way to guarantee the survival of Jewish civilization would be to destabilize an early semiotic model and replace it with another. In an attempt to preserve their status as exclusive encoders, the Sadducees rejected not a derivative corpus of secondary sources but the millions who cared about the Torah so much that they felt compelled to discuss it day and night, offer as many interpretations of it as they could possibly imagine, and record their discussions in an expansive, inclusive, polyphonic textual body that continues to grow and thrive.

Into the future: a new life

Annihilation, both cultural and physical, becomes a clear and present danger with the destruction of Jerusalem by the Romans, the brutal large-scale killing of the Jewish population, deportation, persecution, and the loss of religious and political autonomy. In this context, the type of interpretation semiotics that Petrilli describes (2015: 65) corresponds perfectly to the hermeneutical tradition that the Pharisees develop as an intertextual mechanism against extinction:

> The role of interpretant remains fundamental. Meanings evolve dynamically in open interpretive processes: the greater the degree

> of otherness in the relationship between interpretant sign and interpreted sign, therefore of dialogism, the more interpretation develops in terms of active dialogic response, creative reformulation, inventiveness and critique rather than mere repetition, literal translation, synonymic substitution, identification.

The story of the tension between the priesthood and the opposition is the story of intertextual dialogism. Sign relations are a product of the willingness of the interpreter to make new connections between narrative constituents from past, present, and future texts. The story of Jesus becomes politically meaningful when viewed as a textual development of the tension between Zerubbabel and Joshua. Along these lines, the latter narrative can be seen as a textual development of the clash between Samuel and Saul, which, in turn, is a textual development of the internal rivalry between Moses, head of the priestly family, and Korah, who represents a failed demand for shared governance.

From a Jewish perspective, the story of Jesus also points forward: to the legacy of Yohanan ben Zakkai. If the campaign that Jesus leads against the Temple ends with the temporary triumph of the priests, who extradite their charismatic opponent to the Romans and arrange for his execution, Yohanan ben Zakkai, a Galilean rabbi who challenges the Sadducees in similar ways, is finally able, forty years later, to eliminate the priesthood and form a physical, spiritual, and political substitute for the Temple in Jerusalem: the academy in Yavneh. If Jesus challenges part of the Mosaic tradition, perhaps in an attempt to protect married women from husbands who are eager to divorce them (Matthew 5:31–2; 19:3–9; Mark 10:2–12; Luke 16:18), Yohanan ben Zakkai cancels – officially, permanently, and rather courageously – an old misogynistic law given by Moses and practiced by the priests (Numbers 5:11–31), putting an overdue end to an obvious injustice (*m.Sotah* 9:9). If Jesus encourages his followers to pay taxes to Caesar and refrain from active rebellion against Rome, Yohanan ben Zakkai perfects this strategy when he approaches the Romans and secures their support for his plan to abandon the besieged Jerusalem and offer Jewish communities a chance of cultural resurrection, reconstruction, and survival. If Jesus, somewhat surprisingly, commends the faith and loyalty of a centurion (Matthew 8:5–13; Luke 7:1–9; John 4:46–54), Yohanan ben Zakkai, an insightful visionary, forms an equally unlikely alliance with a higher-ranking, much more influential Roman figure: Vespasian, the emperor-to-be. Just like Jesus, who rises from the grave to complete his mission, Yohanan ben Zakkai, in a dramatic act that saves the nation, pretends to be dead in order to extract himself from the confines of the city. He surrenders to Vespasian, the Roman commander in charge of suppressing the Jewish revolt, and asks for his help in establishing a new scholarly centre to replace the doomed Jerusalem (*b.Gittin* 56a–b; *Lamentations Rabbah* 1:31; *Midrash Proverbs* 15; *Avot de Rabbi Nathan* Recension A 4:30–70; Recension B 6:1–17). Accompanied by a peace-seeking following of devoted pupils and fellow rabbis, he fools the militant zealots who block all escape routes, emerges from his coffin, and leads the people out of an embellished mass grave – the headquarters of an oppressive, self-serving priesthood – and into a new era, a better future. Rather than blind adherence to sacrificial rituals, the future that Yohanan ben Zakkai envisions for the nation is based on a more egalitarian principle: the perpetual reading, expounding, and rewriting of sacred literature. This type of personal and communal engagement with the text is an invitation, extended to every Jew, to join the scriptural history and cultural production of the people. In typological terms, if Jesus is the promise, Yohanan ben Zakkai – whose idea of dialogic hermeneutics gives those who choose to participate in it a new, more meaningful, more complete life – is the fulfilment.

The fact that this particular reading of Jesus and Yohanan ben Zakkai might be unorthodox testifies to the inescapably incomplete, flexible, often idiosyncratic attributes of interpretation semiotics. Flavius Josephus, for example, tells his own version of the dramatic meeting with Vespasian, with himself in the role of the resurrected Jew, coming out of his suicide cave to greet the Roman commander as a future emperor. Some of the aforementioned Talmudic sources are quite critical of Yohanan ben Zakkai, presenting him as a collaborator who should have tried harder to save Jerusalem, while the tenth-century Book of Josippon, which borrows from Josephus and Talmudic literature alike, attributes the cave story, including the surrender to the Romans, to a Jewish war hero who is commended, despite the disapproval of his fellow warriors who prefer death to defeat, as a peace-seeking saviour. Certain voices in modern Israel, especially those that subscribe to nationalistic, militaristic, or geopolitical forms of Judaism, view Yohanan ben Zakkai, once again, as a traitor (Marx 2010), whereas Neusner (1970) tries to offer a more complex biography.

Kadushin (1938: 13–14), in his early analysis of the "grammar" of rabbinical literature, pays attention, in a manner that seems to foreshadow the work of Eco, Sebeok, Deely, Petrilli, and other semioticians, to the inherently indeterminate nature of intertextual sign relations. Eco, in a series of "pre-semiotic" writings published in the 1960s, uses terms like polyvalence, fecundity, multi-interpretability, openness, and infinity to refer to the hermeneutical principles outlined earlier by Kadushin. Ochs (1993: 51), who translates what Kadushin calls the "organic complex" or rabbinical thought into semiotic terms, describes a textual ritual not unlike the one that Sebeok imagines as the best protection from extinction:

> Rabbinic literature of the Talmud has generated the paradigmatic discourse of Jewish communal life. The discourse is paradigmatically midrashic, which means that it presents judgments about everyday conduct as interpretations of the performative implications of Scriptural discourse for life in some contemporary Jewish community. For the rabbinic interpreter, Scriptural texts function as genuine, or incomplete, symbols that acquire definition only in their performative interpretations. Each such interpretation is occasioned by some crisis within the community, which crisis functions as an indexical sign of some inadequacy in the community's rules of conduct. For the rabbinic interpreter, these rules constitute "Jewish tradition" as it is realized in this community at this time. The work of interpretation is to recommend some way of repairing the communal crisis by reforming some aspect of the tradition.

As a manifestation of intertextual autonomy, the self-referential tradition of biblical and rabbinical literature is what Sebeok may have had in mind: the kind of vital, habitual, popular practice that anticipates the disappearance of context and recognizes the volatile, dynamic, and inevitably interrupted nature of sign relations. The most effective substitute for the cryptic rituals that a so-called atomic priesthood might create and disseminate, therefore, is a cumulative, ever-expanding corpus of texts that complement, revise, respond to, elaborate on, or otherwise interact with previous texts. In semiotic terms, if Jesus is a sign, rabbinical literature is a potential signified: a set of textual practices that liberates the people from the limitations of time and place and increases the chances of a more autonomous, more secure, longer-lasting mode of existence.

References

Aichele, George. *The Control of Biblical Meaning: Canon as Semiotic Mechanism*. Harrisburg, PA: Trinity Press International, 2001.

Deely, John. *Basics of Semiotics*. Bloomington: Indiana University Press, 1990.

Deely, John. *Semiotic Animal: A Postmodern Definition of "Human Being" Transcending Patriarchy and Feminism*. South Bend, IN: St. Augustine's Press, 2010.

Eco, Umberto. *The Open Work*. Translated by Anna Cancogni. Cambridge, MA: Harvard University Press, 1989 (originally published in 1962).

Garfield, Susan. "Atomic Priesthood Is Not Nuclear Guardianship: A Critique of Thomas Sebeok's Vision of the Future", *Nuclear Guardianship Forum: On the Responsible Care of Radioactive Materials* 3 (1994): 15.

Houser, Nathan. "Signs and Survival", *The American Journal of Semiotics* 29, no. 1–4 (2013/2007): 1–16.

Kadushin, Max. *Organic Thinking: A Study in Rabbinic Thought.* New York: The Jewish Theological Seminary of America, 1938.

Marx, Dalia. "מיתוס עתיק בשירות ההווה: יציאת רבן יוחנן בן זכאי מירושלים והקמת יבנה" ["An Ancient Myth in the Service of the Present: The Exodus of Yohanan ben Zakkai from Jerusalem and the Establishment of Yavneh"]. *Akdamot* 24: 156–176.

Neusner, Jacob. *A Life of Yohanan ben Zakkai.* Leiden: E.J. Brill, 1970.

Ochs, Peter. "Rabbinic Semiotics." *The American Journal of Semiotics* 10, no. 1–2 (1993): 35–65.

Peirce, Charles S. *The Collected Papers of Charles Sanders Peirce*, Volumes 1–6, edited by Charles Hartshorne and Paul Weiss. Cambridge, MA: Harvard University Press, 1931–1935; Volumes 7–8, edited by Arthur W. Burks. Cambridge, MA: Harvard University Press, 1958 (originally published 1866–1913).

Petrilli, Susan. "Identity Today and the Critical Task of Semioethics." *The American Journal of Semiotics* 31, no.1–2: 55–116, 2015

Proskurin, Sergei G. 2010. *Essays in Contemporary Semiotics.* New York, Toronto, Ottawa: Legas Publishing.

Sebeok, Thomas A. "Pandora's Box in Aftertimes". In *I Think I Am a Verb: More Contributions to the Doctrine of Signs*, 149–73. New York, London: Plenum, 1986, pp. 149–73. A close variant appears (as "Pandora's Box: How and Why to Communicate 10,000 Years into the Future") in *On Signs*, edited by Marshall Blonsky. Baltimore, MD: Johns Hopkins University Press, 1985, 448–66.

West, Donna E. "The Semiosis of Indexical Use: From Degenerate to Genuine." The American Journal of Semiotics 28.3–4 (2012), 301–323.

Wildavsky, Aaron. *The Nursing Father: Moses as a Political Leader.* Originally published in 1984. Page numbers refer to the second edition (published as *Moses as Political Leader*, Jerusalem: Shalem Press, 2005).

8

THE BŪDSHĪSHIYYA'S TOWER OF BABEL

Cultural diversity in a transnational Sufi Order

Marta Dominguez Diaz

Languages are crucial to the ways religious identities are manifested, and they illustrate the dynamics through which religious practice is articulated. Transnational religious organizations span vast territories, incorporating peoples from different cultures who speak different languages. Processes of cultural and religious re-rooting, combined with others of cultural hybridity, emerge as a result of the re-territorialization of religious groups.

In this chapter, I argue that the ways in which languages are used in transnational religions explain the mechanisms and limitations that propitiate and bound the preservation of the cultural character the group possessed before expanding, as well as the emergence of processes of cultural hybridity. The analysis of language usages in transnational religious groups questions the supposed enhancement of truly 'international' religious cultures by providing a more nuanced picture.

Sufism, the mystical dimension of Islam, is socially articulated through Sufi Orders (sing. *ṭarīqa*). These are religious organizations led by a Sufi *shaykh*, a *walī' 'Allāh* ('friend of God'), a person who is believed by his followers to have saintly attributes, to possess a sacred aura (*baraka*)[1] that makes him capable of performing miracles. Rather than regional cults, Sufi Orders today tend to be transnational due to the effect migration has on them but also because they can often be quite successful in attracting new members with cultural backgrounds different from that of the *ṭarīqa*'s leadership. This is the case of the Sufi Order on which this chapter concentrates. The Sufi Order Qādiriyya Būdshīshiyya is an organization founded in 1942 by a Moroccan Berber, Abu Madyan ibn Munawwar al-Būdshish, and led since 1972 by Sīdī Hamza. The *walī* of Berber origin lives in the Order's central lodge (*zāwiya*) in Madāgh, a dormant village in the north-eastern Moroccan province of Berkane, in the Rif, a mainly mountainous area inhabited largely by Berber peoples. The area has been influenced by Spain, since it was part of its protectorate between 1912 and 1956, when Moroccan independence was recognized. There was an interruption of the Spanish presence in the region during the period 1921–26, when the Republic of the Rif, led by the Berber leader Abd El-Krim, was established as a breakaway state after a Berber victory in the Rif War. For further information, see Pennel (1982, 1986). In the eastern part of the Rif is Madāgh, a small village of 2,366 habitants, where the central lodge (*zāwiya*) of the *ṭarīqa* Qādiriyya al-Būdshīshiyya is situated. Located in the administrative region of *l'Oriental*, in north-eastern Morocco, the town is about fifty-two kilometres north of Oujda and sixty kilometres from the North African Spanish enclave of Melilla.

By looking at the use that devotees of the transnational Sufi Order Qādiriyya al-Būdshīshiyya make of languages, this paper discusses the relative capacity and the restrictions for cultural mixing that may occur in transnational religious organizations. More precisely, it suggests that despite the Būdshīshiyya's universalist (i.e. sometimes trans-ethnic, sometimes transnational, and at yet other times transcultural) appeal, religious identities cannot escape their cultural locatedness and are always engrained in the cultural context(s) to which they belong. Such situatedness comes to question the very transnational character of the Order, and it determines the ways in which it negotiates its local/global character.

The Būdshīshiyya – local and transnational

Half a century ago, this Moroccan *ṭarīqa* was very different from what it is today. According to the locals' account, in the past, after an initiatory ritual (*bay'a*), the aspirant (*faqīr*)[2] would become a full-time student of the master at his lodge. Most of these disciples were young Riffian males from the surrounding areas. Despite the fact that the mother tongue of the majority was Riffian Berber, the language of religious instruction was Arabic (*fuṣḥā*). *Fuṣḥā* is the formal and standardized written version of Arabic, and it substantially differs from the spoken versions of Arabic that exist throughout the Arab world. The regional oral varieties are often learned at home and form Arabs' native language, while the formal language is subsequently learned in school.

This is because of the predominant oral usage of Berber dialects, the weak standardization of its writing system,[3] and, notably, the fact that Islamic education has traditionally been conducted in *fuṣḥā*. Not only scholarship but religious matters, more broadly, are generally dealt with in *fuṣḥā* throughout the Arab world, whereas spoken versions of Arabic and other Middle Eastern languages are left for communicating about subjects socially considered of less importance. The sociolinguistic dimensions of this functional diglossia have been analysed by Albirini (2011).

On occasion, the lodge was open for collective celebrations, days in which visitors filled the *zāwiya* for the performance of a collective Sufi ritual (*waẓīfa*)[4] and visited the saint either to get impregnated from his saintly aura or to ask him for 'saintly' intercession (e.g. to cure a man or help a woman to conceive). Although these visitors would hold *waẓīfa* sessions in their own neighbourhoods on a regular basis, and Sufism was definitely intricately embedded in the local culture to which they belong, in their lifestyles and value systems, and it informed societal dynamics, they were not strictly speaking members of the Order: they had never been initiated, and they could not be considered students of the *shaykh*, and in most cases they would not refer to themselves as *fuqarā'*. They constituted, nonetheless, the bulk of Sīdī Hamza's followers and were numerically the most significant group of devotees. Back in those days, both ethnicity and language (labels closely intertwined in the Berber means of self-identification) played an important role, with not only most *fuqarā'*, but also the visitor-devotees – mostly from a Riffian background but from other Berber areas as well – being of Berber origin. Accordingly, diverse dialectal variants of Berber could be heard in the lodge; there appeared to be a diglossic milieu in which *fuṣḥā* was used by the leadership and in rituals, and variants of vernacular Berber were used for informal communication amongst devotees.[5] (Berber is composed of different variations that some consider separate languages, whereas others refer to them as dialects. Speakers of the Riffian variant (Tamaziġt) are estimated to be around 3.4 million, according to the 2004 national census. The language is spoken in the Rif mountainous area and is the less-spoken Berber dialect in Morocco. By contrast, the Tashelhit variant is the most widely spoken, as it covers the whole of the region Souss-Massa-Drâa, as well as the Marrakech-Tensift-El Haouz and Tadla-Azilal regions. The 2004 national census estimated that 4.3 million people spoke Tashelhit, but linguists have raised the estimates to 8–9 million (Stroomer 2008). Other variants include the Senhaja, the Ghomara, and the Figuig-based version of Shilha.)

During the 1970s and coinciding with the change in the leadership of the Order, the *ṭarīqa* initiated a process of profound transformations: doctrinal changes that would allow the organization to surpass its traditional ethnic and linguistic boundaries to embrace followers from other cultural provenances in Morocco and abroad. The by then *tabarrukiyya* character of this *ṭarīqa* – in the language of Sufism, being led by a master who has received 'Godly authorization' to teach a selected group of students how to reach spiritual enlightenment – was incorporated into a *tarbawiyya* ethos or educational enterprise. That meant that the Order adopted a proselytizing attitude by which it broadened the scope of discipleship – a selected group of Riffian *fuqarā'* was transformed into a much more numerous following from further away. In doctrinal terms, this missionary approach was justified by arguing that the transmission of spiritual knowledge previously occurring face-to-face was no longer needed, since the new approach considered spiritual love (*maḥabba*), a spiritual state that emanates from the *shaykh* towards his disciples, the central means for the acquisition. Such emanation is not physical, and therefore the corporeal presence of the *shaykh* was seen as unnecessary for the transmission of skills that would make the spiritual enlightenment of the disciple possible. In terms of language, it also meant that, since religious learning was no longer acquired though verbal communication, leader and disciple did not necessarily need to know the same languages. Followers of other cultural provenances could then be incorporated. Overall, it was thought that this doctrinal shift that the new leader of the Order, Sīdī Hamza, legitimized in the eyes of his followers enabled the transformation of a regional cult into the transnational religious organization that it is today.

As a result of this, the Order soon came to grow exponentially, by attracting members in Morocco and abroad. Notwithstanding this, today the Būdshīshiyya still attracts the majority of its followers from the Berber-speaking regions of the country, despite the burgeoning number of devotees who are urban Moroccans and whose mother tongue is Dārija (the spoken colloquial Maghrebi variant of Arabic), with some even coming from the more privileged French-speaking classes. The Order has moved beyond the Alauite country and now has groups of followers in Western Europe, the Americas (North and South), the Gulf, south-east Asia and sub-Saharan Africa. Despite this global representation, the more sizeable and stable following outside Morocco is located in France, Belgium, Spain, the United Kingdom, and the United States, and it is largely connected to the economic Moroccan diaspora. There is, nevertheless, a growing number of followers without a Maghrebi background, both those of Muslim origin and new converts to Islam. This translates into an impressive amalgam of languages spoken within the same religious organization; sometimes at the Būdshīshiyya international gatherings, one can hear all the variants of Berber and dialectal Arabic, alongside a variety of European languages (e.g. English, French, Spanish, and Catalan). One may wonder, then, what effect this 'Tower of Babel' has had on the internal dynamics of the Order: whether language plays a role in enhancing or impairing the development of a truly transnational Būdshīshiyya character. In other words, how do languages and religious identities interact within this *ṭarīqa*?

Loosening Būdshīshiyya's Berber identity?

In looking at the interplay between religious identity and language in the Būdshīshiyya, we first need to consider the role played by Berber in the transnational organization. Berber is the original 'mother tongue' of the Būdshīshiyya, and it is still the language of most of the devotees and the leadership of the *ṭarīqa* alike. Yet, as the Order grows, significant numbers of followers, as well as influential characters close to the *shaykh*, have nothing to do with this Berber character. Most of the urban devotees of the *ṭarīqa* in Morocco, as well as all of its international following, have no knowledge of Berber, and neither is it spoken by influential personalities within the Order, such as Faouzi Skali.[6] Berber does not even figure as being part of the character of the Order. When European followers describe what they see as the inclusive, cosmopolitan, and multicultural character of the Būdshīshiyya, with a mixture of cultures being mentioned, the Berber identity is rarely mentioned.

Several factors are related to this seemingly intentional omission of *la Berbérité*. On one hand, in Morocco, Berbers have been portrayed both in the past by colonial officials and in the present by a significant proportion of religious scholars and anthropologists as the bearers of folk Islam, of heterodox religiosities that partly maintain their pre-Islamic character. Many of the rituals associated with these religious lifestyles are Sufi – for example, the exaltation of saint veneration and tomb visitation and the performance of sacrifice rituals connected to the agricultural calendar. The category of 'folk Islam' is often contraposed with 'official' and more orthopraxical versions of Islam. Although it is true that urban religious scholars in Morocco have, over the centuries, criticized the exalted nature of such Sufi performances as falling outside Islam, the clear-cut dichotomy between rural/urban, official/folk, and great/little Islamic traditions that was made famous by Gellner (1969), and has been reproduced ever since in much anthropological literature, seems to be less representative of reality than is often assumed (Chachoua 2002; Silverstein 2012).

The turn of the Būdshīshiyya towards a more sober understanding of Sufism seems to initially have been a response to the criticisms exerted from the traditional centres of religious scholarship. But it also coincides with the approximation of the leadership of this Order with the centres of political power in the country. Thence, the Būdshīshiyya is part nowadays of a political project, one that fundamentally affects the very character of the Order. Morocco's minister of religious affairs, Ahmed Taoufik, is a declared devotee of the organization, as other high-placed government officials certainly are. This move towards a less ecstatic understanding of Islam tries to present itself as being in tune with the traditional bastions of Moroccan religious scholarship – namely, Islamic Scripturalism and a respect for Maliki sharia law. If the Baraka-driven exalted nature of a ritual-oriented religiosity is too commonly associated with Berber Islam, this Scripturalist trend is more often connoted and, although misleadingly, identified with the Arabic character of religious scholarship in Islam – Berber regions were not entirely alien to centres of religious scholarship of this more normative kind, and representatives of this more 'orthodox' version of Islam have also existed in Berber areas past and present (Munson 1993; Silverstein 2012).

The *makhzen*'s[7] support for the Būdshīshiyya is a mechanism for trying to counteract the modernizing approach to Islam proposed by the Islamists, in particular those of the Justice and Charity Party, but not by them alone.[8] In trying to counteract so-called Islamism by adopting a more sober, normative leaning, the Būdshīshiyya ends up detaching itself from the more folkloristic expressions of Sufism characteristic of the Berber and rural regions of the country. Non-Berber followers and even the leadership of the Order will, as a result, be sympathetic to publicly taking part in Sufi ritual practices that have traditionally been seen as the core of *maraboutic* ('folk') Berber Islam, yet all these practices have been purportedly deprived from its more intense, ecstatic character, to a certain extent tuned down and turned, one may argue, into commoditized expressions of religious emotionality that occur in a rather contained manner. The diminishing use of Berber dialects in this *ṭarīqa*, particularly by the authorities and most prominent members, coincides with this broader religious 'de-Berberization' of Moroccan Sufism, strongly affected by the aforementioned political agenda. The Berber character of the Order is mentioned neither in most of the Būdshīshiyya's written materials, nor by members in most of the non-Berber locations. Considering that Berber identity has turned into a very heated and debated issue within Moroccan politics,[9] it is not surprising that a Sufi Order with such a public political profile within the country would prefer to be seen neither as supporting nor as rejecting *la Berbérité* or any form of Berber activism.

The Būdshīshiyya's tower of Babel

What else can the ways in which languages are used by adepts in this religious organization tell us about the internal functioning of the Order? First, to the larger extent that the Order is still a mainly Moroccan organization, the *ṭarīqa* is a vivid image of the multilingualism that is quotidian in the North African country. Berber- and Dārija-speaking devotees, especially the younger and middle aged, are generally good connoisseurs of French and *fuṣḥā*, the languages of education and administration. Besides, Berbers can often speak Dārija, although less often can speakers of Maghrebi Arabic communicate in Berber. Those followers from the Rif are sometimes fluent in Spanish, due to the geographical proximity, the broadcasting of Spanish TV and radio stations in Northern Morocco, the historical influence of the Protectorate, and the fact that some of them work on the other side of border, in the Spanish enclave of Melilla.[10] Some of the urban, better-off members of the Order consider French to be their mother tongue, although they can also speak both *fuṣḥā* and Dārija. Few Moroccan members are good English speakers. By contrast, and despite the fact that most Europeans are taught a second language in school, the fluency of languages is significantly more limited among them than among their Moroccan counterparts. European devotees without a Moroccan background can, in most cases, fluently communicate only in their mother tongue (whether Catalan and/or Spanish, French, or English), with very few being able to speak another European language, let alone Berber or Arabic. This monolingualism critically affects communication between members – because devotees tend only to commingle with others who speak their very same language.

Partly the result of this tower of Babel, although also influenced by different cultural approaches and ways of living Islam, devotees of Sīdī Hamza do not gather all together. Each of the cities in which the *shaykh* has followers contains several small congregations, groups that only occasionally meet with one another. The Order is thus made of a large network of small groups, each of them holding separate weekly gatherings. In terms of relating to *fuqarā'* from other countries, the monolingual character of many devotees makes communication between these congregations very difficult. Thus, despite the facilities for communication brought about by the internet and social media, devotees with different mother tongues rarely communicate with one another. Most of the followers meet only in international gatherings held by the Order annually, where congregations from different countries have the opportunity to meet to perform devotional rites together. Not being fluent speakers of a second language means that when members meet, their interaction is limited; most of the Berber members I encountered, for example, know that there are foreign followers of Sīdī Hamza yet had never spoken to any of them. Similarly, in most instances, members of the various European locations have spoken only on a very few occasions to devotees from other cultural provenances. Language plays a significant role in those interactions – Belgian and French followers tend to meet more frequently with one another than with Spanish or British ones, for example. The few European devotees that know more than one language establish long-term relationships with devotees from other countries and maintain these relationships through occasional visits and by email and social media; as for the rest of them, who constitute the majority of followers of the Order, the degree to which they taste the transcultural character of the transnational religious organization is rather minimal.

Migration has definitely played a very significant role in the spread and consolidation of the Būdshīshiyya beyond its original milieu: it has helped in the building of communities in which people care about each other in circumstances in which the traditional networks of friendship and family are diluted due to the reality of living away from home. Thence, the Būdshīshiyya is particularly strong among Moroccan students studying at university, as well as among economic Moroccan migrants to large European cities – not only those of poorer backgrounds in search of unqualified jobs, but also those of wealthier backgrounds looking to further develop their careers. These weekly gatherings help people feel less alone in difficult circumstances, find company, and help one another search for a place to live or a job. For many of these migrants, these weekly gatherings become pivotal loci both of community life and, especially among the communities that have migrated to Europe, of cultural preservation. For that reason, perhaps, Būdshīshiyya groups of Moroccan migrants in Europe tend to be less open to strangers than other groups; their almost exclusive use of Arabic or Berber evidences the importance that the religious organization acquires for these members in terms of preservation of cultural identity.

In France, there are French-speaking groups whose interaction with Moroccan *fuqarā'* living in France is rather limited, and in which a significant proportion of members are second-generation Moroccans raised in France and with a very different way of understanding their religious identity (e.g. they tend to be less conservative, frequently choosing not to wear a headscarf). The fact that these second-generation Moroccans decide to gather with the French-speaking group can be seen as a symbol of cultural assimilationism or as a larger degree of cultural adaptation. Overall, the sum of these factors seems to impede a more fluid construction of a transnational character, a sense of identity shared by all members of the organization.

Communication between the different groups occurs only at a (hierarchically speaking) higher level. Those who have been authorized by the Order's leadership to 'lead' and organize a local congregation are in regular contact with the central lodge and the leaders of other congregations. It seems that in these exchanges French, rather than Berber or Arabic, has come to be the vehicular language. French, a language originally used in this Order as a proselytizing means – that is, to approach new audiences in Europe – has acquired a new dimension within the Būdshīshiyya's linguistic ecology, turning out to be a marker of identity and a commonality shared by a significant number of followers and leaders, and the language in which most of the written materials by its members are produced.

This chapter contends that since most of the exchanges occurring transnationally among congregations do so in a top-down manner, such communication is not effective in building a sense of communion and amity amongst followers in different locations. At the same time, the rather limited spontaneously motivated interactions between followers from different countries are an obstacle to developing similar understandings of religion. On the contrary, a myriad of very different cultural identities are preserved, with markedly different ways of living Islam. To that extent, arguments advanced by scholars predicting that globalization would inevitably blur traditionally held signifiers such as ethnicity, language, and nationality at the expense of new forms of social identification – by religious belonging, for example (e.g. Nederveen Pieterse 2004; Tomlinson 1999) – are shown to be misleading. In the case of the Būdshīshiyya, those traditional identifiers are still the basis on which members of the Order define themselves and relate to others. They still are the driving forces that inform the ways in which relations within the organization are manifested. Thus, transnational religious organizations like the Būdshīshiyya, despite defying contextualization with its global expansion and universalist, proselytizing appeal, are still framed within the social and cultural forces that determine the milieus in which they operate.

Literary output and identity

The lack of identity cohesiveness among the followers of Sīdī Hamza is also evident in the Order's written output. The Būdshīshiyya's official website seems to organize and present similar material in different ways in different languages, evidencing differences in the ways devotees from diverse geographical contexts perceive and experience religion. For example, the fact that there is no Berber version of the official website may once again point towards the widely oral nature of the language, its disenfranchisement from a more formal side to religious matters, and the leadership's effort not to be 'encapsulated' into a Berber identity that could make it hard to attract members from other ethnic backgrounds in Morocco. The front page of the website, in Arabic, instead introduces some of the names contained in the Order's *silsila*,[11] the central source of religious legitimation of a Sufi *shaykh* which is based on genealogy. It also gives a moral-cum-theological justification for the 'method' Sīdī Hamza uses to instruct his *fuqarā'*. This sort of explanation can make sense only for people already familiar with the social composition of Sufism – people for whom Sufism is just part of the society's religious fabric – and may, as a result, build comparisons between this and other Orders and their different 'techniques' for transmitting spiritual knowledge to devotees.

The front page of the Spanish-language site, by contrast, needs to provide a broader introduction to key Sufi concepts (e.g. *sirr*,[12] *silsila*) and even to recognize the mere existence of Sufi Orders. It reads: "there are all around the Muslim World, but also in the rest of the world as well, numerous Sufi brotherhoods that are named after their founders, such as the Shadhiliyya, Tijaniyya, Naqshbandiyya, Qadiriyya, et cetera".[13]

Although some of the *silsila* names appear as well, the text is of only an introductory nature, confirming that it seeks to address a Spanish-speaking readership that can visit the site – namely, potential converts to Islam interested in joining this religious organization. For French readers of the website, there is no reference to the *silsila*, and there are no introductory concepts. There is only a quote by a Sufi master and the Qur'anic verse 10:99–100 translated into French – in itself, one can argue, indicative of a progressive decreasing sacral character attributed to Arabic.[14] Similarly, the American site[15] presents Qur'anic verses translated into English, with no reference to the original Arabic. Additionally, the British site contains a saying attributed to ʻAbd al-Qādir al-Jīlānī, the founder of the Qādiriyya tradition to which the Būdshīshiyya pertains, also translated into English and with no reference to the Arabic original text.

Other digital materials produced by members of the Order also offer a window into the amalgam of cultures and related variegated understandings of Islam that exist within this religious organization. In seeking to attract different types of followers, these texts address issues in rather different ways. By looking at the websites produced in different countries, one can grasp nuances that are specific of each country's cultural and religious milieu and the contextual circumstances in which the Order exists. For example, a poetic language full of literary metaphors common in medieval Sufi literature is used on the American website. This can be interpreted as a reference to a way of understanding Sufism typical of the West. In the United States, for example, a translation of the works of the medieval Sufi poet Rumi has been the best-selling book of poetry in translation for more than a decade.[16] Partly due to this editorial success, many Americans associate Sufism with poetry, and although literary output has been an important facet of Sufism throughout history, poetry represents only a tiny part of the complex and multidimensional religious phenomenon that is Sufism. The reference to poetry is thus illustrative of these particular circumstances.

The British case illustrates other dynamics; one of the UK websites captures the religious hotchpotch characteristic of contemporary British multiculturalism in terms that would certainly be impermissible to many Moroccan Muslims: "We welcome readers from all faiths, and none, to explore the pages on this humble website."[17]

Another website in English is quite indicative of the community orientation of the Būdshīshiyya in the United Kingdom, featuring subsections on employment, activities organized by the Order, etc. on the home page, with less attention being paid to the more formal religious aspects of Sufi legitimation (*silsila*, etc.).[18] One of the French-language websites, *Saveurs Soufies*, a magazine on Sufism produced by the *ṭarīqa*, is more centred on behavioural and ethical aspects of Sufism, featuring several articles on Sufi doctrine, accompanied by notes on the origins of the Order and with textual comments and reviews on Moroccan Sufi treatises.

Decreasing the 'holy' status of Arabic?

When scrutinizing an Islamic religiosity from the perspective of analysing its language usages, an understanding of the role played by Arabic is essential. For, being an expression of Islam, the sacral status of *fuṣḥā* in this organization plays an undisputed role, although the transnational expanding of this Order beyond the predominantly Muslim cultural milieu that is Morocco into places in which the culture of Islam is that of a minority may tend to imply that such hegemony could be (even if unconsciously) somehow negotiated. There are some incipient signs that may point in that direction. In the Būdshīshiyya, Arabic is the language of ritual praxis, and it is generally considered by its members as an adequate vehicle for accessing the divine. Some see it as a unique means of communicating with *Allāh*; others prefer to think of it as just one of many existing languages that might be used for that purpose. Nonetheless, even those of the latter opinion argue in favour of adopting a sense of 'religious consistency' (i.e. if you are a member of a Sufi Order, you should stick to the 'Muslim mechanisms' and not adopt an eclectic stance of also following some aspects of other religions). This is a claim particularly addressed to some Western audience members who could be defined as having a somehow 'New Age' approach to religion and who might undertake religious practices from many different sources, including the practice of *dhikr*, and thus seem to be potentially interested in joining this *ṭarīqa*.[19]

In general, however, while it is prescribed by the leadership of the Order, the use of Arabic in ritual praxis is neither questioned nor challenged, dissimilar to what happens in other Sufi organizations that have taken root in Western Europe and North America. For example, the leader of the largely Western-based Sufi organization of Pir Vilayat Khan is an example of the decreasing relevance of Arabic in transnational groups that have sprung up outside the Muslim world. The leader of this group can code-switch from French to Dutch to German to English with fluidity, more fluently than he can use any Islamic language. An introductory scrutiny of the use of languages in some transnational Sufi Orders has been provided in Hermansen's (2006) analysis of their literary output.

Although Arabic is respected as a liturgical language, the sacrality of Arabic as typically understood in the Muslim world (i.e. the idea that its knowledge is spiritually enlightening and that it is a language in which religious scholarship shall be conducted) seems to be less common among members of this *ṭarīqa* in the West. Even if the majority of converts who join the Order adopt Arabic names after being initiated, it is surprising to see how few are the non-Arab members who have learned Arabic in the course of the years they have been Sīdī Hamza's *fuqarā'*. In terms of knowledge of Arabic, the Order seems to be divided: those who were Muslims before joining the Būdshīshiyya see the knowledge of Arabic as an important prerequisite of religious correctness and piety. Converts, on the other hand, do not make such a close association between Arabic and their religious devotion – coinciding with a trend also observable among other non-Būdshīshī converts to Islam in Europe.[20]

It is not only that they do not incorporate the sense of 'holiness' that born Muslims tend to attach to Arabic (in that regard, we shall note that even Berber members attach to Arabic a notion of religious purity); it is also that the fact of their being members of a Moroccan Sufi Order fails to motivate the willing to learn either Berber, Dārija or *fuṣḥā*. In fact, and accordingly, their interest in knowing the culture and society from which their *shaykh* comes tends to be very limited. Most of the convert followers with whom I have spoken know little about Morocco and do not seem to intend to learn any of the languages used in the country. In terms of the lack of interest in learning *fuṣḥā*, one may see the lodge in Birmingham as an exception, since they hold weekly courses of Arabic for followers and non-followers alike, yet a very substantial proportion of the devotees in this location are not converts to Islam. (A significant proportion of Sīdī Hamza's British devotees are Muslims of South Asian origin, with some prior knowledge of Sufism, although in its South Asian forms (which in some aspects can be quite different from North African religious culture). They are 'revert' Muslims, people born in Muslim families who decide to join a branch or understanding of Islam different from that of their upbringing. Further analysis on Muslim reversion can be found in Gilliat-Ray (1999). Offering courses of Arabic when lodges are placed in predominantly Muslim neighbourhoods may also be a means of attracting Muslim members to the Order.[21] Overall, and more broadly, the scarce interest in *fuṣḥā* held by converts can be equated to a similar dilution experienced by the status of *fuṣḥā* when Islam has encountered non-Arab populations. Although the sacral role is mostly preserved in these cases, the use of other languages is introduced when dealing with religious scholarship and other aspects of religious life.[22]

Not seeing *fuṣḥā* as a central element of their religious apprehension is thus concomitant to the transnational expansion of the Order, and it represents a shift from what used to happen among *fuqarā'* in the Būdshīshiyya prior to the 1970s. Related to that, the decreasing centrality of *fuṣḥā* among some followers may be seen as a consequence of moving away from the *tabarrukiyya* orientation, a more religiously scholarly oriented approach, and adopting a less text-based stance. Although a formal initiatory ritual (*bay'a*) is still required in order to become a *faqīr*, no compulsory reading and/or writing is involved in the religious duties of the post-1970s devotee. *Fuṣḥā*, a language considered not only liturgical in Islam, but also central in the development of Islamic scholarship, was mandatory for students of the *shaykh* in the past, but it is no longer a requirement. In contrast to the dedicated religious learning that characterised membership in the Order before the 1970s, only weekly attendance at the collective ritual *waẓīfa* sessions and 'morally adequate' social behaviour are considered requisites of the Sīdī Hamza's *faqīr* nowadays.[23] In this sense, the clear-cut distinction between visitors and devoted *fuqarā'* that existed prior to 1970 has, to a certain extent, blurred. Although there are marked differences between the religious identities of members from different locations, the degree of commitment seems to have levelled out. Most of Sīdī Hamza's followers nowadays have a similar degree of commitment; they perform weekly ritual sessions and pay an occasional visit to the *shaykh*'s lodge on a more or less regular basis, but the *shaykh* no longer knows who has and who has not been formally initiated. The embodiment of religion has come to replace a more rationalistic approach in this *ṭarīqa*. Accordingly, little relation to the written and to a discursive dimension is retained. Thence, not only are devotees not particularly dedicated to the study of religious sciences (*'Ulūm al-dīn*) and of any scriptural Islamic tradition, but they mainly concentrate on the adequate performance of rituals. Sīdī Hamza himself also adopts a non-discursive, more corporeal approach to religious authority.

Thus, he may be seen in presence of the followers, be touched by them, and be venerated by them, but very rarely will he speak to the public; he does not give speeches, he does not publish books, and in general he does not address his admirers in verbal ways. There is only a 'set' of sayings, phrases attributed to him, that circulate among devotees in various languages, but this is a short text of a rather informal nature (i.e. it has never been published, and it is not clear when was it produced or how was it compiled). Not wanting to leave a significant body of literature to his followers is consonant with the emphasis on a more experiential approach to religion characteristic of the Order. Sīdī Hamza prompts *fuqarā'* to "experience" Sufism instead of "reading" about Sufism. The intellectual and the experiential are often portrayed as antithetical approaches to pursuing the spiritual quest, the intellectual being commonly portrayed as an obstacle to the attainment of the experiential. Accordingly, his sayings assert that:

> Understanding is not acquired through books. It would be too easy to lower oneself and collect all the books written about Sufism in order to acquire such understanding. True knowledge comes to you from inside, from your heart, only the heart can truly understand that nothing is outside God.

Besides, a justification of the non-verbal character of the relationship between devotee and *shaykh* is provided:

> He who understands the value of the shaykh knows that his relationship with him does not require exchange of words. You see me. I see you and that is ample and bountiful. [. . .] Verbal teaching is not necessary. It is only the transformation of hearts which is important. Sidi Boumadiane, Sidi Hamza's own master and shaykh, only rarely spoke.[24]

The emphasis on the experiential and the turning away from a verbally transmitted and/or text-reliant understanding of Islam has in fact given this Order its capacity to surpass cultural and linguistic barriers more easily. Getting involved in the intricacies characteristic of *'ulum al-dīn* would imply a degree of dedication and an extent of cultural estrangement (by immersing oneself into a religious culture that is completely alien) that imply a degree of commitment most followers probably would not wish to undertake. Yet, if we see this as a mechanism that is able to enhance the success of religious proselytization and speed up its pace, it is yet to be seen how the resulting loosening of a unified character and the mushrooming of a multiplicity of *fuqarā'* identities may play out in keeping a like-minded understanding of Islam throughout the *ṭarīqa* and ultimately in keeping a cohesive organization running.

The Būdshīshiyya as a speech community: enhancing a sense of belonging

All the aforementioned instances (e.g. the lack or very minimal existence of a textual religious culture and the lack of communication between different language groups) seem to point towards a tendency in which territorial expansion could jeopardize the cohesiveness of this religious organization. By incorporating a myriad of religious identities with little connection to one another – namely, the local Berber groups – the mainly Arab groups from Morocco's cities, the various North-American and European strands, and the Moroccan diaspora communities in Europe and the Americas, the Būdshīshiyya increases its cultural diversity at the risk of weakening the internal solidity of the *ṭarīqa*. However, there are also indicators of how particular usages of language in this transnational organization actually contribute to developing a sense of sharedness among devotees from different cultural backgrounds and how a yet-incipient feeling of camaraderie among members from seemingly dissimilar backgrounds is enhanced by the particular usages of words. Despite not being able to communicate with one another, I have noted, Sīdī Hamza's followers from different origins use the same Arabic terminology when referring to religious issues. The ways in which these phrases and terms are used by devotees of this organization make the Būdshīshiyya, I would contend, a 'speech community',[25] in the sense of its being a group of people who share a series of understandings and expectations regarding the use of a particular vocabulary; a glossary worth noticing is in *fuṣḥā*.

Accordingly, most members, regardless of their cultural provenance, use the same terminology to refer to certain things that are considered pivotal in their religious ecology. For example, weekly ritual sessions are known as *waẓīfa*; the chanting and reciting of litanies that constitutes them are known as *dhikr*; the stage one enters when in ecstasy while in a *waẓīfa* is known as *ḥāl*; the formal initiatory ritual for officially joining the Order is known as *bay'a*; fellow followers are referred to as *faqīrā* (fem.), *faqīr* (masc.), and *fuqarā'* (pl.); and God, of course, is always *Allāh*. This religious terminology is never translated; people speak their other languages but refer to those concepts by using the term in *fuṣḥā*. Also, the mastering of such a glossary varies from one member to another; thus, those who have been members for longer tend to use such religious terminology more fluidly and more often.[26]

The use of this vocabulary is a way of identifying devotees of Sīdī Hamza, and it develops a transnational speech community that provides a sense of a shared identity in an otherwise extremely variegated group. The employment of this terminology, I argue, is central for boosting amity among members, as it turns out to be a common source of easily discernible identification. Despite their not being able to converse fluidly, knowing that these terms pervade their cultural and linguistic differences definitely strengthens individuals' sense of belonging to the Order. Within this, the degree of adequateness and appropriateness in the use of this lexicon by each devotee is viewed as a mark of the member's degree of commitment to the organization and devotion to Sīdī Hamza. Being less accurate and proficient in the use of such vocabulary signals a more recent incorporation to the group, and a fluid use of such vocabulary implies having a longer trajectory in the Order and a subsequent higher 'mastery' of religious knowledge. The usage of this terminology when referring to religious matters, thus, manifests a two-fold mechanism of enhancing the *fuqarā*'s belonging to the Būdshīshiyya: (a) it tells outsiders that the person belongs to the religious group and (b) it informs insiders about the nature and relative duration of the allegiance between the devotee and the *ṭarīqa*. Thus one may assess a devotee's status within the religious system by looking at this form of 'terminological dysphemism'. Overall, this is a symbol within the Būdshīshiyya of how religious identity is dependent upon a use of language: in the same ways that the lack of a common language can contribute to disconnection, the sharing of a vocabulary can support processes of reconnection.

Concluding remarks

Both religion and languages are instrumental in shaping ideologies and are potent vehicles for the articulation of identities. The interplay between the two is worth exploring, for, as they are intricately related, each can help us to understand the other. Languages, this chapter has shown, are key contributors to the ways in which religious communities operate and are pivotal determinants of the internal dynamics of transnational religious organizations. The plural map of religious lifestyles that is characteristic of this *ṭarīqa*, the Būdshīshiyya, is a result of its territorial expansion. This transnational spread is manifested in the myriad of languages used nowadays within the Order. The paper envisages this linguistic multiplicity as an obstacle to communication between members from different linguistic backgrounds, in what I have metaphorically equated with a 'tower of Babel' that tends to weaken the former internal solidity of the organization.

Other challenges as a result of the transnational drive include the preservation of the sacral status of *fuṣḥā* beyond its mere ritual function. Members outside Morocco show little interest in learning the language, as many of them (particularly converts to Islam) do not conceptualize a connection between sacrality and *fuṣḥā* as something exclusive. Convincing them to approach *fuṣḥā* in this rather more sacral manner may turn out to be arduous, especially because the Order does not promote a primarily scholarly based religious identity among devotees, and Islamic scholarship could actually be a way of bringing devotees to the learning of the Semitic language. In terms of the (lack of) interest in *fuṣḥā*, the analysis of language in this organization has elucidated differences in understanding religion between religious converts and reverts. European devotees who are born Muslims share this understanding of the sacral nature of Arabic with their Moroccan counterparts, dissimilar to what occurs among converts, who either see it as only one of many languages that can be used to access the divine or do not make this linkage at all.

There are, however, some signs that seem to indicate the preservation of a united sense of belonging and a shared religious identity amongst members of the Order worldwide. In exposing those indicators, scrutinizing languages has proved to be significantly fruitful. In particular, the usage of a shared terminology when referring to religious issues gives Sīdī Hamza's devotees a sense of uniqueness – of differentiation from other Muslims, from other Sufis, and from other 'New Age' seekers. The usage of a common glossary thus enables us to look at the Būdshīshiyya as a 'speech community', and the identification of such 'community' can accordingly be conceived of as evidence of the existence of patterns of religious cohesiveness within the Order. These trends may serve to glue together the organization, while other trends contribute to the mushrooming of religious identities and the growing disenfranchisement between them.

Overall, language seems to play a key role in the signifying of religious identity in this Sufi Order. The avenues and obstacles to mutual camaraderie lead us to conclude that transnational organizations like the Būdshīshiyya can be only partially understood as instances of religious and cultural cross-pollination. A globalized *ṭarīqa* such as this is a meaningful example of the complexity of religious and cultural hybridization, and studying it by looking at the ways in which languages are used gives us new considerations to contemplate in analysing how identities are regulated in transnational religious groups.

Notes

1 *Baraka* (Ar.): generally translated as 'blessing'. It refers to the spiritual potency that holy individuals, places, and/or objects are believed to have. Since its existence is thought to be tangible, it is believed that *baraka* can be transmitted to those who come into contact with the person or thing that possesses it. Sufi leaders are generally believed to have this attribute, which makes them, in the eyes of their followers, capable of performing miracles.

2 *Faqīr/a*; pl. *fuqarā'/faqīrāt*: in Arabic it means 'poor', and in popular parlance it is used for a homeless person, a pauper, or a beggar. Among North African Sufis, the term is used as a synonym for 'disciple'. The traditional connotations of the term describe someone who entirely accepts the will of God and has no private property, considered to be indispensable attributes of the *faqīr*. Today, reformed Sufi orders such as the Būdshīshiyya use the term simply to refer to a member of the Order, irrespective of his/her degree of spiritual commitment.

3 The original alphabet in which Berber was written is known as the Tifinagh writing system. In 2003, a modernized version of this alphabet used by Berbers before the switch to Arabic script and then Latin characters was made official in Morocco. Nevertheless, and despite attempts to promote it, its

usage is still quite limited, resulting in Berber languages being still largely oral, and written use of Latin characters is more common.

4 *Waẓīfa*; pl. *waẓīfāt*: litany consisting of various formulas chanted and/or recited in repetitive ways. These rituals are more often known as *dhikr* (remembrance), since a substantial part of such sessions are dedicated to invoking *'Allāh*, in a commonly held belief that God's essence will come to the scene.

5 The use of two distinct languages for different purposes in a given context is also common in other religions, particularly in religions for which one particular language retains a sacred status. In the case of the Abrahamic religions, this 'holiness' is ascribed to particular languages that are believed to be those of Divine Revelation and of the Sacred Scriptures. In societal milieus in which those religions operate, the use of the 'holy' language is often maintained for liturgical purposes, while another vernacular is used by the community for communicative means (Fishman 2002; Liddicoat 2012).

6 Skali, not a member of the Būdshīshiyya family but certainly someone with influence within the Order, received an anthropology degree from the Sorbonne with a thesis on Moroccan Sufism. His Wikipedia site is indicative of what he represents for many Būdshīshiyya devotees: '*ecrivain francophone, il se situe entre l'Orient et l'Occident et œuvre pour le dialogue des hommes et des cultures*'. See 'Faouzi Skali – Wikipédia', https://fr.wikipedia.org/wiki/Faouzi_Skali, last accessed on 26 August 2016. Although he has no bearing on the authority structure in a traditional sense (since he is unrelated by blood to Sīdī Hamza), he is respected and followed by members and often mentioned by them as an inspirational source. Sometimes his name is praised and mentioned with more recurrence than that of the leading *shaykh*. Skali embodies a modern approach to Sufism, advocating a cosmopolitan form of religious life. He is a public figure: the director general of the Spirit of Fes Foundation and the famous festival of 'sacred' music hosted at the city, frequently appearing, also, in the media and having authored several books on Sufism (e.g. Skali 1985, 1996, 1999, 2004).

7 *Makhzen* is a term used in Morocco to refer to the government and its institutions; it centres on the king and his notables, as well as on a selected body of high-ranking officials.

8 An interesting exploration of the relationship between Islamists and the state can be found in Daadaoui (2011) and Zeghal (2008). The (until recently) most relevant Islamist leader Morocco has had so far, 'Abd al-Salām Yāssīn, who passed away in 2012, was a former member of the Sufi Order who pushed to become leader of the organization at the same time Sīdī Hamza did in 1972. He was expelled from the Order and created his own group, which within years turned out to be the most prominent Islamist political party in the country. His Berber origin is, equally, almost never mentioned, although a strong bastion of support to this group was initially geographically focused on the Rif region, the region where they won their first elections and constituencies. This very same region is the one from which the Būdshīshiyya initially gained most of its followers and the only place where it has had political representation; members of the Būdshīshiyya family have run as independent candidates in Berkane and have long been involved in local politics. How Berber activism has played

out within the Sufi/Islamist dyad that has recently developed in the country is a hitherto under-researched, yet very interesting, subject.

9 For an analysis of the actual state of affairs between the Berber movement and the state, see Maddy-Weitzman (2013).

10 Melilla's *souqs* ('markets'), where smuggled products are sold at cheap prices, are an important resource for the economy on both sides of the border. Contraband makes a substantial contribution to the Riffian economy and is central to the survival of the enclaves of Ceuta and Melilla. Every day at the El Barrio Chino pedestrianized frontier, hundreds of people, mostly women, transport smuggled goods. It is estimated that from Monday to Thursday around eight thousand *porteadoras* enter Melilla. McMurray (2001) has written an interesting ethnography about the social dimensions of smuggling in the region.

11 Most Sufi Orders claim that their leadership is connected by blood to that of the Prophet Muhammad; this spiritual lineage, the chain of authority that traces the genealogy of a Sufi Order back to the Prophet of Islam, is known as the *silsila* and constitutes the main source of religious authentication.

12 The *sirr* is the secret formula that is believed to be transmitted by God to the *shaykh*. It gives the *shaykh* details on how to lead his disciples in the different stages within the spiritual path they have to pass to achieve ultimate Godly Realization.

13 My own translation of the original in Spanish in a website from the Order: '*Existen por todo el mundo musulmán, pero también alo largo del mundo entero, numerosas cofradías sufíes, que llevanel nombre de sus fundadores como la Shadhiliyya, la Tijaniyya, la Naqshbandiyya, la Qadiriyya, etc.*'; see 'Sheij Sidi Hamza – Tariqa.fr', http://tariqa.fr/espanol/index.php, accessed 21 August 2016.

14 An elucidating study on how translating the Qur'an into other languages can lead to the weakening/strengthening of the 'holy' status of Arabic in Islam, by the study of a translation into Hausa, can be found in Brenner (1985).

15 See 'Sidihamza.us', www.sidihamza.us/, accessed 10 March 2013.

16 See Jalāl al-Dīn Rūmī, translated by Barks (1995).

17 Available online at 'The Sufi Way | It is quite simple, we love each other, we love GOD, and GOD loves us', https://thesufiway.co.uk/, accessed 21 June 2016.

18 Available online at 'The Zawiya | Home of the Tariqa Qadiriyya Budshishiyya', http://thezawiya.co.uk/, accessed 3 July 2016.

19 For a preliminary analysis of New Age religiosities among members of the Būdshīshiyya, see Haenni and Voix (2007). I have also discussed this subject in my doctoral work (Dominguez Diaz 2010).

20 For example, Zebiri, in her study of converts to Islam in the United Kingdom, found out that only four of her interviewees had a knowledge of Arabic that was sufficient to allow them to read original sources (2008: 50).

21 Poston's (1992) study of Islamic proselytization in the West is still one of the most accurate studies done on the subject. Arabic-language courses are common activities with missionary intent.

22 Examples of multilingualism in religious contexts in the case of Southern Africa can be found in Mumisa (2002).

23 A comparative analysis of the devotees' religious identities in different Būdshīshiyya enclaves is the research I undertook as a doctoral study, which was revised and published as a monograph. For further information, see Dominguez Diaz (2010, 2014).

24 Quotes available online at 'Sayings | The Sufi Way', https://thesufiway.co.uk/the-path/sayings/, accessed 21 August 2016.

25 A 'speech community' can be defined as a group of people who share a particular set of rules for language use through interacting together or by sharing certain worldviews and ideologies. Such groups can be geographically bounded (e.g. villages or countries), ideologically or occupationally driven (e.g. political parties or professional communities), or simply groups with shared interests, hobbies, or lifestyles. Members of speech communities use similar speech styles and genres, sets of vocabulary, and grammatical conventions. For further analysis, see Yule (2014) and Patrick (2008).

26 For example, sometimes the plural of *faqīr* will be formed *faqīrs*, which follows the rule for plurals in French, English, and Spanish, rather than the correct Arabic form, *fuqarā*.

References

Albirini, Abdulkafi. "The Sociolinguistic Functions of Codeswitching between Standard Arabic and Dialectal Arabic", *Language in Society* 40 (2011): 537–62.

Brenner, Louis. "The Role of Language in West African Islam", *Africa: Journal of the International African Institute* 55 (1985): 432–46.

Daadaoui, Mohamed. *Moroccan Monarchy and the Islamist Challenge: Maintaining Makhzen Power*. New York: Palgrave Macmillan, 2011.

Dominguez Diaz, Marta. "Revisiting Moroccan Sufism and Re-Islamisizing Secular Audiences: Female Religious Narratives in the *ṭarīqa Qādiriyya Būdshīshiyya* in Morocco and Western Europe Today." Unpublished PhD thesis. London: University of London, School of Oriental and African Studies, 2010.

Dominguez Diaz, Marta. *Women in Sufism: Female Religiosities in a Transnational Order*. London: Routledge, 2014.

Fishman, Joshua A. "Holy Languages in the Context of Societal Bilingualism", *Contributions to the Sociology of Language* 87 (2002): 15–24.

Gellner, Ernest. *Saints of the Atlas*. Chicago: University of Chicago Press, 1969.

Gilliat-Ray, Sophie. "Rediscovering Islam: A Muslim Journey of Faith". In *Religious Conversion: Contemporary Practices and Controversies*, edited by Christopher Lamb and M. Darroll Bryant, 315–32. London: Cassell, 1999.

Haenni, Patrick, and Voix, Raphael. "God by All Means . . . Eclectic Faith and Sufi Resurgence among the Moroccan Bourgeoisie". In *Sufism and the 'Modern' in Islam*, edited by Martin van Bruinessen and Julia D. Howell, 241–56. London: Tauris, 2007.

Hermansen, Marcia. "Literary Productions of Western Sufi Movements". In *Sufism in the West*, edited by Jamal Malik and John Hinnells, 28–48. London: Routledge Curzon, 2006.

Liddicoat, Anthony J. "Language Planning as an Element of Religious Practice", *Current Issues in Language Planning* 13 (2012): 121–44.

Maddy-Weitzman, Bruce. "The Amazigh Factor: State-Movement Relations under Mohammed VI". In *Contemporary Morocco: State, Politics and Society Under Mohammed VI*, edited by Bruce Maddy-Weitzman and Daniel Zisenwine, 109–24. London: Routledge, 2013.

McMurray, David A. *In and Out of Morocco: Smuggling and Migration in a Frontier Boomtown*. Minneapolis: University of Minnesota Press, 2001.

Mumisa, Michael. "Islam and Proselytism in South Africa and Malawi", *Journal of Muslim Minority Affairs* 22/2 (2002): 275–98.

Munson, Henry. *Religion and Power in Morocco*. New Haven and London: Yale University Press, 1993.

Nederveen Pieterse, Jan. *Globalization and Culture: Global Mélange*. Oxford: Rowman and Littlefield, 2004.

Patrick, Peter. "The Speech Community". In *The Handbook of Language Variation and Change*, edited by J.K. Chambers, Peter Trughill, and Nathelie Schilling, 573–97. Malden, MA: Blackwell Publishers, 2002.

Pennell, C.R. "Ideology and Practical Politics: A Case Study of the Rif War in Morocco, 1921–1926", *International Journal of Middle East Studies* 14, no. 1 (1982): 19–33.

Pennell, C.R. *A Country with a Government and a Flag: The Rif War in Morocco, 1921–1926*. Wisbech: Middle East and North African Studies Press, 1986.

Poston, Larry. *Islamic da'wah in the West: Muslim Missionary Activity and the Dynamics of Conversion to Islam*. New York: Oxford University Press, 1992.

Rūmī, Jalāl al-Dīn. *The Essential Rumi*. Translated by Coleman Barks. San Francisco: Harper, 1995.

Silverstein, Paul A. "In the Name of Culture: Berber Activism and the Material Politics of 'Popular Islam' in Southeastern Morocco", *Material Religion* 8 (2012): 330–53.

Skali, Faouzi. *La Voie Soufie*. Paris: Albin Michel, 1985.

Skali, Faouzi. *Traces de Lumière: Paroles Initiatiques Soufies*. Paris: Albin Michel, 1996.

Skali, Faouzi. *Le Face à Face des Coeurs: le Soufisme Aujourd'hui*. Gordes: Editions du Relié, 1999.

Skali, Faouzi. *Jésus Dans la Tradition Soufie*. Paris: Albin Michel, 2004.

Stroomer, Harry. "Three Tashelhiyt Berber Texts from the Arsene Roux Archives", *Studies in Slavic and General Linguistics* 33 (2008): 389–97.

Tomlinson, John. *Globalization and Culture*. Cambridge: Polity Press, 1999.

Yule, George. *The Study of Language*. Cambridge: Cambridge University Press, 2014.

Zebiri, Katherine. *British Muslim Converts: Choosing Alternative Lives*. Oxford: Oneworld, 2008.

Zeghal, Malika. *Islamism in Morocco: Religion, Authoritarianism, and Electoral Politics*. Princeton, NJ: Markus Wiener Publishers, 2008.

Part III

POLITICAL USES OF RELIGIOUS LANGUAGES

9

RELIGION FOR NATION? CHURCHES' LANGUAGE POLICIES IN BELARUS

Nelly Bekus

Belarus is one of the post-Soviet countries which experienced the whole complex of processes associated with becoming a sovereign state after the 1991 fall of the USSR and trying to overcome the legacy of socialism in political, social, and cultural life. The twin revivals of the nation and the church constituted an indivisible part of this process.

After seven decades of suppression of religious life in the Soviet state, churches received a chance to establish themselves as legitimate participants in the transformation of post-Soviet societies. They needed to restore their organizational structures, to reclaim their place in the institutional order of transforming societies, and to restore the people's option of having a religious world outlook. At the same time, the nation-building projects launched by the post-Soviet states faced an urgent need for basing the consolidation of people within the new collective frameworks, and religious tradition could supply an appropriate resource for these purposes.

This chapter explores complex correlations between the project of nation-building and language policies of the Catholic and Orthodox Churches in post-Soviet Belarus. It examines how these churches' vision of Belarusian identity, both national and religious, predetermined their place in the religious landscape of the country.

The weakness of religious arguments in the political struggle in Belarus

From the perspective of ethno-cultural understanding of nation, both language and religion constitute important elements of national ideology that are able either to facilitate or to prevent the emergence of a shared national identity. The nation is a "type of identity whose meaning and priority is presupposed by such cultural factors as an ideology, a language, a mythology, symbolism and consciousness that has achieved global resonance" (Smith 1991: 91–2). Such a definition of 'nation' refers to the major symbolic and cultural elements of a national group's shared heritage, which belong to a subjective order as they focus on the perceptions, memories, beliefs, and values of individuals and communities. Such uniting beliefs and values are defined, among other factors, by religious faith. At the same time, for many constructivist students of nationalism – for B. Anderson, E. Hobsbawm, E. Gellner, and others – the progress of nationalism was intimately bound up with the secularization of social life and the emergence of a secularized middle-class intelligentsia. Nations as new "imagined communities" were completely different from the pre-modern communities of the past. The new mode of self-perception was spread primarily through the various secular channels of informing and influencing society. The national community was a new type of unity which came to being after, and partly because, religion started to lose its importance. The fact that national movements in some countries – such as Ireland and Poland– were developed under the influence of the Catholic Church and were controlled by traditional forces made them, according to Hobsbawm, "non-progressive" (Hobsbawm 1962: 170–1). As Anderson has suggested, rational secularism in the eighteenth and early nineteenth centuries brought with it its own modern darkness, for which nationalism acted as an antidote (Anderson 1983: 17–21).

Nationalism is not so much an ideology similar to other ideologies – like liberalism, conservatism, or even socialism – but should be described as a new integrative belief system which, to a certain extent, replaces religion (MacLaughlin 2001: 98). This dichotomy between religion and nationalism, however, came to be perceived as an attribute of the Western discourse of modernity (Veer 1994). Even within Europe, patterns of modernization and its impact on people's religiosity varies. Instead of a depersonalized Western type of modernity in which religion is pushed to the side, in Eastern Europe, as Brian Porter-Szűcs noted, "we see adherents remaking (and yes, in many contexts abandoning) their religious traditions and institutions as they struggle to make sense of a changing world" (2010: 5). Sometimes, common religion became a factor facilitating a shared sense of national belonging, alongside a common history, shared geography or territory, a common culture, and a common language (Anderson 1983: 6–7). In their searches for historical justification, nationalist movements often take advantage of identifying themselves with the strongest carrier of the national tradition, the Church.

After the fall of the state socialism in Eastern Europe, there was an urgent demand for a new moral basis which would shape collective self-consciousness and manage social disorientation. Religious faith was one of the ideological systems which had not been tainted by association with socialism and, at the same time, represented historical continuity and cultural tradition. In some countries – such as Poland – the Catholic Church was involved in both the anti-communist underground and the country's national post-communist revival. This resulted in the formation of a specific type of "moral majority" (in Poland, Slovenia, and Croatia) which conceives Christian values as the ideological 'cement' of the nation (Salecl 1996: 418).

Belarus constituted a rare exception to this Eastern European rule during the first years of its independence. The nation's ideology, which underpinned the rebirth of the Belarusian nation and political liberalization in the early 1990s, had never directly appealed to a religious tradition to support the collective identity of Belarusians. The reason for this can be found in several factors entangled in Belarusian's political and religious history.

On one hand, the high level of atheization of post-Soviet Belarusian society made religious appeal a rather weak argument in the political debates of the 1990s. Belarus, just like many other former Soviet republics, had experienced not forty but seventy years of brutal atheistic propaganda, accompanied by oppression of priests and believers. A survey conducted in 1989 revealed that 65 per cent of respondents defined themselves as atheists; in 1994, their number fell to 32 per cent. At the same time, only 22 per cent claimed to be believers in 1989, but their number increased to 43.4 per cent[1] in 1994 (Tsvilik 2005: 308–9). In this context, politicians could hardly rely upon religious feeling in seeking instruments for mobilizing the society.

The second essential factor weakening the role of the religious argument in the Belarusian revival was related to the strong historical association of the two main religions, Orthodoxy and Catholicism, with Russia and Poland. Once religion re-entered public life, the cultural association with religious traditions influenced from outside was re-established.

The third factor weakening the power of religious issues in national movements was related to the civilizational role played by Orthodox Church. In the context of ethnicization of religion in the Russian Federation, belonging to Orthodox tradition turned out to be a strong argument in favour of the thesis about Belarus belonging to a Russian cultural domain. This idea, however, was largely contested, if not disapproved, by the ideologists of the Belarusian revival, who put all their energy into re-imagining and re-inventing Belarus as a part of Europe.

Post-Soviet Belarus came to be predominantly culturally Orthodox, though the Belarusian state equally recognizes the main religious holidays of both the Orthodox and Catholic Churches. Sociological studies conducted in 1999 showed that more than 50 per cent of the Belarusian people acknowledged the importance of religion and considered religion an essential factor of the contemporary social development. And more than 80 per cent of believers identified themselves with the Orthodox Church, compared to about 14 per cent who identified as Catholic and 2 per cent who identified as Protestant (Tsvilik 2005: 309).

Churches as agents of external domination

Throughout the history of Belarusian lands, starting with the Polish Lithuanian Commonwealth[2] and later in the Russian Empire, two major branches of Christianity present on Belarusian lands, the Orthodox and Catholic Churches, were instruments of external cultural and political influence in Belarus. Eastern Orthodoxy linked Belarusians with the Russian cultural universe, identified either subconsciously or explicitly as "the Russian faith," while Catholicism has often been seen as the Polish creed connecting Belarusian Catholics with their western neighbour. During the first years of Belarusian independence, these clichés inherited from history were reanimated; however, these strong ethno-religious correlations have become gradually overshadowed by the more nuanced geopolitics of the Churches. While the Orthodox Church in Belarus continues to be a factor unifying the Belarusian cultural tradition with Russia, several symbolic steps were undertaken in order to institutionalise the separate status of the Belarusian Orthodox Church, albeit under the supervision of the Moscow patriarchate.

In 1989, the Bishops' Council of the Russian Orthodox Church (held on October 9–11, 1989) adopted a definition of the formation of the Belarusian exarchate of the Moscow patriarchate, with the official name of the "Belarusian Orthodox Church." Yet, compared to the Ukrainian Church of the Moscow patriarchate, the Belarusian Orthodox Church has much less autonomy. For the first time in history, however, Belarusian Orthodoxy was granted the official institutional frameworks coinciding with the borders of the Belarusian state. Institutional infrastructure of the Orthodox Church within the country also was formed at that time: Gomel eparchy was restored, and a new Brest eparchy was established (in January 1990.) In May 1992, eparchies in Turov and Vitebsk were established. There were 609 parishes in the republic by the end of 1991. By 1 July 1994, the Belarusian Orthodox Church comprised 850 parishes and 8 monasteries, and by the time of the second visit of the Patriarch of Moscow Aleksiy II to Belarus in 1995, there were already 918 parishes, 3 monasteries, and 6 convents. Minsk Theological College (closed in 1963) started its work again in 1989; further on, Belarusian Theological Academy was set up in accordance with the 1993 resolution of the Synod.

At that time, the Vatican also created administrative structures for Roman Catholics in the former USSR, and among these were first three dioceses in Belarus – Minsk-Mohilev, Pinsk, and Grodno; Vitebsk diocese was created in 1999.[3] In 1991 Kazimir Świątek was appointed the first archbishop of the new archdiocese of Minsk-Mohilev; in 1994 he became the first Cardinal in post-communist Belarus. He remained the head of the Belarusian Catholic Church in the years 1991–2006, and under his lead the policy of Belarusisation of Catholicism in Belarus was initiated immediately after the country became independent.

The history of the activity of two main Christian-denomination churches in Belarusian lands and the way in which they influenced Belarusian national development displays a paradoxical interdependence between politics and religion. Religion was often instrumental in establishing political domination over the territory, while changes in political power were accompanied by shifts in national religious affiliation. Similarly, in the past the Belarusian language often became a victim of the political choices made by the Belarusian elites. At the same time, language was a major factor in churches' popularization of religion.

The history of Christianity in Belarus starts in 992, the year when Polatsak's Orthodox bishopric emerged. Several years later, in 1005, a second bishopric was founded in Turau. The Byzantine version of Christianity became the state religion in Belarusian lands, which at that time were part of Kievan Rus, and was spread throughout the realm by force of decree (Zaprudnik 1993: 14). The introduction of Orthodox Christianity to the lands of Rus predetermined the future of East Slavic civilizational development: it would henceforth be based on the Byzantine cultural tradition and opposed to the European tradition of Western Christianity.

Christianization brought to the principality of Polatsak literacy, education, and fine arts. Located on the border between Rus and Western Christianity and maintaining independent ties with both Byzantium and Europe, Polatsak became

> a major cultural centre with its own original schools of architecture, which were infused with the best traditions of the culture of the contemporary world, with monasteries, where chronicles were composed as was done in Kiev, Novgorod, and other major centers of Rus
>
> (Alekseyev 1966: 220)

Until the thirteenth century, there were two Belarusian Orthodox bishoprics, and they were part of the Kievan archdiocese. Both of them always remained under the strong influence of the centre of Russian Orthodoxy (first in Kiev, later in Moscow).

In the fourteenth century, Belarusian lands, together with Lithuanian ones, were consolidated in the Grand Duchy of Lithuania (GDL). Some Belarusian historians consider this state as a largely Belarusian one. It was Old Belarusian which became the language of the ducal chancellery, courts, chronicles, and various official translations in GDL. The cultural domination of Belarusian in GDL was determined in the first place by Christianity. As Zaprudnik (1993: 19) wrote:

> The Christianity and literacy that had been practiced in the Belarusan lands for almost four centuries before the Christianization of Lithuania ensured the prominent role of the Belarusan territories within the Duchy, where local customs were well preserved and the language of which had gained official status and was used even in Lithuanian ethnic areas.

Alongside the ideas of the Reformation coming to Belarus from Western Europe, Old Belarusian language penetrated canonical religious texts at the beginning of the sixteenth century. At that time, numerous Belarusian magnates adopted Calvinism. Its teachings brought religious postulates to real life and oriented its adepts towards a more active understanding of religious practice, which enabled Calvinism to play a positive role in the cultural development of Belarusians. At the beginning of the sixteenth century, Calvinism gained adherence among businessmen, tradespeople, and artisans in many towns. Reformation evoked lively debate and a flood of polemical publications; numerous printing houses contributed significantly to the development of the Belarusian literary language, which would eventually become the basis for the modern Belarusian revival after the "twilight zone" of the eighteenth and first half of the nineteenth century (Zaprudnik 1993: 34). Belarusian historian Stanislau Akinchyts called the Reformation the golden age of Belarusian history, during which the humanist ideas of the Renaissance largely influenced the development of Belarusian culture (2001: 122). The development of printing houses and numerous schools in local religious communities, characteristic of the Calvinist epoch in Belarusian history, was conducive to the increasing popularity of the Belarusian language in public and cultural life. The major problem of that period was a growing cultural and religious division between the nobility and lower classes. Magnates accepted Calvinist ideas with great enthusiasm, which can be partly explained by their wish to reduce the economic privileges and power of the Catholic and Orthodox clergies. At the same time, the vast majority of townsfolk and rural inhabitants remained Orthodox. Moreover, as Zaprudnik wrote, "the attachment to the faith of their forebears was seen by a majority of Belarusans as a matter of national survival" (1993: 34). In this context, even the use of the Belarusian language in Protestant publications, education, and religious services, prompted undoubtedly by patriotic motivations on the part of many, "was widely viewed as an instrument used by the upper class to advance its own economic and political goals" (Zaprudnik 1993: 34). Due to its devotion to the ideas of the Reformation, the gentry were viewed by the common people as an agent of alien cultural and political influence.

This ideological conflict between two directions of influence on Belarusian development constituted the main feature of that period. Western cultural and religious influence, brought together with Calvinist ideas to Belarus and even clothed in Belarusian language, was opposed by the Orthodox faith, which linked Belarusians with the East and made them feel a part of the Eastern Slavic cultural and spiritual tradition. Also, Muscovy had often used the religious pretext of "reuniting Eastern Christians" under the banner of Orthodoxy in its conflicts with the Grand Duchy and went to war against GDL "for the primordial Russian lands". Orthodox Christianity being a religion of the Belarusians was, in fact, a major argument proving their "Russianness".

The decline of the Reformation in Belarusian lands was related not only to the active Catholic counteraction to Protestantism, but also to the wars that the Grand Duchy of Lithuania waged against Muscovy. The wars caused large numbers of casualties amongst town-dwellers – the most active section of the population during the Reformation – and reduced the number of nobles who adhered to Luther's and Calvin's ideas. As a result of the Lublin Union between GDL and Poland, signed in 1569, a new Commonwealth was created.

As part of the overall Counter-Reformation strategy and as a way to weaken the power of Russian Orthodoxy in Belarusian lands, a religious union between Orthodoxy and Catholicism was signed and the Greek-Catholic Uniate Church was established in 1596. The new church preserved the Orthodox rites while recognizing the supreme authority of the pope and accepting some fundamental dogmas of the Catholic Church. Such a union "was much desired as a unifying factor in a state where vast eastern territories shared the same East Christian rite with the covetous Muscovy" (Zaprudnik 1993: 35).

The evaluation of this religious compromise between the two branches of Christianity in the context of Belarusian development remains ambiguous even today. Those who see the essence of Belarusianness as belonging to Orthodox Christianity view the Uniate Church in a negative light: it was instrumental in the Polonization and Catholicization of Belarus and led to the alienation of Belarusians from Russia (Chigrinov 2004). Historians who see Belarus as part of the Western European tradition emphasize that the Uniate Church preserved many aspects of Orthodox religious rituals, combining them with the use of the Belarusian language in education and sermons (Arlou and Saganovich 1999). By 1839, when the Uniate Church in Belarus was forcibly converted to Russian Orthodoxy after the incorporation of the territory into the Russian empire, more than 80 per cent of the Belarusian peasantry belonged to it. Just as two centuries before, when the Uniate Church was introduced as a 'national denomination' for Belarusians, this time the return of the Orthodox Church was presented as a 'way back to their genuine religious tradition'. And both were set up by force.

Surprisingly, at the beginning of Russian rule the authorities allowed the Belarusian gentry to drift towards Polish language and culture. Polonization affected mainly the higher strata of the population, and by the time of the Belarusian national revival in the late nineteenth and early twentieth centuries, the major part of the Belarusian common folk was again Orthodox. In this way, the Belarusian nation-to-be was deprived of the opportunity to develop its "own" autochthonous high culture, while the return to the Orthodox faith "removed the religious basis for a popular notion of a distinct Belarusian nation" (Snyder 2003: 45). The absence of a unitary religious foundation to Belarusian national determination led to the enhanced role of the only alternative source of self-differentiation – the Belarusian language. The language, however, was a rather weak argument in the rhetoric of national awakening at that time. In the nineteenth century, Belarusian was considered the speech of honest folk.

It was an "uncodified, low-status Slavic dialect located morphologically between Polish and Russian, whose speakers were located socially between Polish culture and Russia power" (Snyder 2003: 41). The history of the Belarusian language embodies the unprecedented story of the rise and fall of an extensive and important literary language which was lost and developed again on a different basis (Waring 1980: 129). During the eighteenth and nineteenth centuries, education in Belarus was conducted in Polish, Latin, or Russian, and the churches used Latin or Church Slavonic. The Belarusian language gradually lost its status as a "language" and started to be viewed as a mere dialect. After the 1863 uprising, Belarusians could not publish in Belarusian in the Russian empire. At that time, Russian policy regarding Belarusian lands was aimed at a complete elimination of a distinct Belarusian idea and the transformation of the Belarusian lands into western Russian ones. According to the concept of so-called *Zapadnorusizm* – "Western Russism" – Belarusians were considered a regional group of the great Russian nation.

The Belarusian national revival movement in the late nineteenth and early twentieth centuries was an almost classical example of small-nation nationalism in Miroslav Hroch's terms (1985). Belarusians were dominated by a ruling class of more or less alien nationality of either Russian or Polish origin and throughout their history had lived in a state shared with other ethnic groups and with no continuous literary tradition. The initial stage of the national awakening of Belarusians started with the activity of writers and poets who disseminated an awareness of the linguistic, cultural, and social ideas of Belarusians as a distinctive ethnic group. Meanwhile there was not even certainty as to whether Latin or Cyrillic was the proper alphabet for Belarusian. This uncertainty was in a way a symbolic one and replicated the historical hesitation of Belarusians between the East and the West. Belarusian publications appeared in both Roman and Cyrillic scripts. Publishing in Cyrillic made it possible to reach those Belarusians who could read Russian, while using the Latin script (and often Polish spelling) served to create distance from Russia.

What was truly characteristic of the first wave of Belarusian nationalism was the combination of socialist revolutionary ideas with the ideas of national revival. This combination led to a specific filtration of the cultural memory of Belarusians. Maurice Halbwachs has written about the difference between the memories of different social classes. In his view, different social strata behave differently in the arts and in life with regard to the content and the values of the cultural tradition, because they have different values themselves (1969: 357). The cultural project of the Belarusian nation, formulated by Belarusian activists in the late nineteenth century, was based on the memory of the lower social strata, which inevitably led to a filtration of the Belarusian cultural tradition. Language became one of the main instruments and, at the same time, a victim of this filtration. The Belarusian revival was built upon the repudiation of the 'old' literary language (and with it, of all that had been created in that language). To a great extent, this was connected with the revolutionary trends of the time and dictated by the spirit of the struggle for social equality.

The resurgence of language was part of the national awakening and was expected to become an instrument of revolutionary agitation. At that time, religion had practically lost its influence on Belarusian development. The Uniate Church had ceased to exist in Belarusian territory, the Catholic Church operated in Latin and Polish, and the Orthodox Church used Church Slavonic. Moreover, the Orthodox faith was one of the factors which enhanced the idea of Belarus's cultural and historical commonality with Russia, while Belarusian national awareness was supposed to be built in opposition to this unity. Meanwhile, Polish-speaking Roman Catholic clergymen became active participants in the Belarusian nationalist movement. This, however, created additional problems in the popularization of nationalist ideas: the idea of national self-determination in opposition to Orthodox Russia, as formulated by those who were considered 'others', was not easily accepted by the masses.

Thus, until the end of the twentieth century, when Belarus became independent and religion re-entered the public sphere after seventy years of atheistic state propaganda, the development of the Belarusian language took place entirely independent of the church and religion.

The twentieth century: language and religion on separate paths

Further nationalization of Belarusian life continued under Bolshevik rule, after the Belarusian Republic was created in January 1919. By granting Belarusians their own republic, the Bolsheviks in fact agreed to recognize the Belarusians as a distinct nation. Nicholas Vakar stated that the founding of this republic served a double purpose: (1) to attract into and maintain within the Soviet system those elements of the population to whom the communist idea might not otherwise appeal and (2) to integrate their nationalism with the world revolutionary forces (1956: 139).

The overall outcome of the Soviet period for Belarusian history was, generally speaking, Russification. Under Soviet rule, however, Russification was not a goal in itself for Soviet ideologists. "[I]t would be more accurate to speak about Sovietization – the formation of a specific Soviet culture, which the Russian language served, rather than Russification" (Yekadumau 2003: 186–7).

The policy of Russification was officially introduced under Stalin's authority in the late 1930s. However, this had been preceded by a decade of intense nationalization of Belarusian life. The Belarusian nationalists, after a short period of hesitation, accepted the creation of the BSSR. Many of them held leading posts in the national government. In a way, the aspirations of Belarusian nationalists achieved their end in this republic. Here the task of cultural enlightenment, education, and forming a national consciousness was carried out with the help of the state, based on its institutions and often tied to the socialist ideology which rejected religion and old bourgeois traditions.

The Belarusian government declared a hard line in the realization of the nationalization of public life: the Belarusian nationality was declared predominant, and Belarusian became the official language. A system of obligatory Belarusian courses was created in education, and the return of the Belarusian nationalists was encouraged. The BSSR reported to the USSR

Soviet of Nationalities in 1925 that in the central agencies of the republic only 26.9 per cent of the employees spoke Belarusian. In 1928 the number had risen to 80 per cent. The press became almost exclusively Belarusian. "The attitude to the Belarusian language and culture changed considerably. Belarusian was no longer a language of peasants but the language of the state" (Mironowicz 2004). In the education system, by 1930 93.8 per cent of schools, compared to 28.4 per cent in 1925, operated in the Belarusian language. The result of the first period of Belarusian nation-building within the Soviet state was "the remarkable success" of Belarusisation (Martin 2001: 261).

Soviet ideology did not leave its citizens any space for a religious outlook. The church was forced to the margins of public life. At the same time, in eastern Poland,[4] where many Belarusians still lived, the Orthodox Church had a chance to become a national denomination for Orthodox Belarusians. With the support of the Polish authorities, the Autocephaly of the Orthodox Church in Poland was approved by the Byzantine patriarch in 1924. However, the major language used by this congregation was Church Slavonic, while in non-liturgical communication with believers Polish gradually started to prevail. The attempt in 1921 of the Belarusian national committee to make Belarusian the liturgical language in the Orthodox Church in this area, at the Congress of Orthodox clergy and religious activists of the Lithuanian eparchy, failed. In 1922 the Synod of the Orthodox Church allowed the use of languages other than Russian in services, but this was an opening not only for Belarusian but also for the Polish language.

Belarusian national organizations repeatedly asked church authorities to start using Belarusian in worship and homily, as well as in the religious education of Belarusian children and in the seminary in Vilna. There was only one parish church where this was allowed by Archbishop Feodosii (Mironowicz 2008: 318). At the Congress of Orthodox Belarusians in 1930 in Vilna, the Central Belarusian Church Committee was established. Two magazines were published by this committee: *Svetach Belarusi* (*The Light-Bearer*) and *Golas Pravaslaunaga Belarusa* (*The Voice of the Orthodox Belarusian*). These publications became the apogee of the Belarusisation of the Orthodox Church in Poland. However, lack of support from the state, church authorities, and the clergy resulted in the gradual decline and marginalization of this process. The Orthodox Church was fully dependent on the Polish government and soon became a tool for the Polonization of Belarusians (Yazykovich 1992: 281).

Attempts to establish the Autocephaly of the Belarusian Orthodox Church with Belarusian as a working language were again made under German occupation in the years 1942–4. Aware of the Moscow patriarchate's negative attitude to the idea of the Belarusian Orthodox Church's Autocephaly, religious activists used the occasion provided by occupation to declare their independence. The main argument behind their aspirations was the idea that Christianity came to Belarus not from Moscow but directly from Byzantium, and therefore Belarusians had the right to a national ecclesiastical institution equal in rights with the Russian Church. Several years later, in emigration, this church declared unification with the Russian Orthodox Church Abroad. The Belarusian Orthodox Church Abroad re-emerged again in the 1950s, in cooperation with the Ukrainian Church. Parishes were registered in Germany, Great Britain, the USA and Australia. The liturgical language in this Church is Belarusian. The church supports the cultural, educational, religious, and scientific activity of the Belarusian diaspora (Yazykovich 1992: 282–3).

Orthodox religious tradition in Belarusian state-building

The Church's rebirth after 1991 took place amidst the wholesale reconfiguration – political, social, and cultural – of public life in the country. Belarus became a sovereign state and had to find new ideological foundations for the national community. Being predominantly an Orthodox country in terms of cultural tradition, Belarus found itself in the paradoxical situation in which its religion, which could eventually provide an additional basis for the nation's cohesion and unity, was prescribed the role of a "branch" of a grand narrative of the Orthodox civilization dominated by and centred in Russia. "The new nationalists" who entered the political scene in the late 1980s and the beginning of the 1990s developed their vision of the Belarusian idea, which was opposed to Soviet and Russian political and cultural "domination". National ideology had the appearance of a liberation struggle triggered by dependence of the Soviet state dominated by Russian culture and language. The project of nationalization activated in the early 1990s aimed at reducing the effects of Russification.

The government formulated a long-term plan, which scheduled the Belarusisation of the educational system, the media, and public life. In 1992, the Deputy Minister of Education, Vasil Strazhau, foretold that in ten years' time the entire Belarusian system of education would shift to using the Belarusian language. The Belarusian language appeared in educational institutions, on television, and on the radio, and all this amounted to a common process of leaving Russia and "returning to everything Belarusian." In this context, religious tradition shared with Russia was one of the most evident arguments at odds with the nationalization policy. The revival of the Orthodox Church and a return to this religious tradition as an alternative to a Soviet ideology did not help to distance Belarus from Russia. On the contrary, the popularity of the Orthodox religious tradition furthered the claim that the essence of the Belarusian idea was being a part of the common with Russian cultural and civilizational space.

Official policy overtly recognized the privileged position of the Orthodox Church in Belarus. A bilateral agreement, sometimes referred to as a "concordat", between the Belarusian exarchate of the Russian Orthodox Church and the Belarusian state was signed in 2003. It is difficult to discern whether or how the exarchate encouraged the government to include a reference to its traditionalism and grant it an exclusive position amongst religious denominations, but it was clear that the Orthodox Church appreciated the message this piece of legislation sent (Richters 2012: 132). According to the agreement between the BOC and the Belarusian state, the primary spheres of cooperation between them related to education, preserving the historical and cultural heritage, health care, social care, humanitarian aid, supporting family life, working with the military service, environmental issues, etc.[5]

The agreement between the BOC and the Belarusian state claimed that it did not intend to infringe on the rights of any religious denomination or Belarusian citizens. It also stated that the BOC and the Belarusian state would work together in their "action against pseudo-religious structures that represent a danger to the person and society".[6] This agreement became the legal and moral foundation of the Orthodox Church's social and public activity. The state recognized the Orthodox Church as an important social institution whose historical experience, spiritual potential, and centuries-long cultural tradition had played a defining role in the religious, cultural, and state tradition of the Belarusian people. In addition, the agreement stated that "the religious and cultural values preserved by the Church constitute the historical legacy of the Belarusian people and serve as the foundation of its national idea" (Kotliarov 2004: 160).

The role of the Orthodox Church as a traditional Belarusian faith is not reflected in its language policy. For a long time, its official language was exclusively Russian, while Church Slavonic was used in liturgy. In 1992 Metropolitan Mikolai, of the Belarusian Orthodox Church Abroad, visited Belarus and met with the exarch of Belarus, Metropolitan Filaret. In itself, such a meeting was an unprecedented event, as the Moscow patriarchate, to which the Belarusian exarchate was subordinate, was strongly opposed to the idea of a Belarusian national church and considered it non-canonical. During the meeting, though, no canonical or dogmatic differences between the two churches were found. However, the language issue – namely the use of the Belarusian language in religious services – was determined to be a major barrier to the unification of the two churches.

At the same time, the standpoint of the Orthodox Church in Belarus is not univocal.

In 1992, Metropolitan Filaret sanctioned and blessed the activity of the main national-minded community of the BOC, the Belarusian Orthodox Brotherhood of the Three Martyrs of Vilna[7] (led by protopope G. Latushka), which was founded in 1992 at the Church of Saints Peter and Paul in Minsk.[8]

This brotherhood initiated the process of the introduction of a national dimension into the religious life of Orthodox Belarusians. The Brotherhood worked to record and to publish the history of the Orthodox Church in Belarus as well as religious literature in the Belarusian language. Their projects included such compendiums as "Shrines of Belarus", "Saints of Belarus", "Parishes and Monasteries of the Belarusian Orthodox Church", and the "Belarusian Orthodox Calendar". One of their main activities related to translating and publishing religious literature in Belarusian. Together with the Orthodox Brotherhood of Saints Cyril and Methodius in Białystok (founded in 1989), the Brotherhood of the Three Martyrs of Vilna launched the Orthodox periodical entitled *Pravaslaue* (*Orthodoxy*).[9] The preface to the first issue mentioned a desire to promote the unity "of all Orthodox Belarusians of the world based on national values and the Holy Orthodox Church".[10]

According to the ideas preached by the Brotherhood, the Orthodox faith is a genuine Belarusian tradition, and the Russian roots of this tradition have no particular meaning in the present religious life of Belarus. For example, particular emphasis has been placed on the fact that major prayers in the Prayer Book (in Belarusian, published by the Brotherhood) have been translated from the original Greek and not from Church Slavonic language.[11]

In the more than twenty years since 1992, when the Brotherhood was founded, the status and symbolic weight of this community in the Belarusian Orthodox Church has not developed. It was founded as and remains a cultural and research centre uniting representatives of the national intelligentsia and religious and cultural activists who are both national-minded and Orthodox. At the same time, institutionally, its members continue to be a part of the Belarusian Orthodox Church of Moscow patriarchate. Studying the Belarusian Orthodox tradition outside of Russian religious and cultural influence, they have never questioned belonging of the Belarusian Orthodox Church to the Moscow patriarchate. This is why the Brotherhood's activity has been criticized from the nationalist perspective as "Belarusian-speaking, but Russian in its essence." By being part of the BOC, it re-inforces the legitimacy of the Russian Orthodox Church and supports the idea that Russian Orthodoxy has not been "anti-Belarusian" (Adamkovich 2009).

While the general context of the Brotherhood's activity has been described in terms of strengthening the national spirit of the Orthodox faith in Belarus, the historical framework of the Belarusian idea, which its members represent, appears to differ essentially from the national idea, articulated and promoted by the Belarusian political and cultural opposition (Bekus 2016: 79–80).

The Brotherhood actively propagates the idea of use of Belarusian in liturgy, and its efforts have resulted in several churches across the country periodically serving liturgy in Belarusian. The first liturgy in Belarusian, in Minsk, took place in 1994 in the parish of Our Lady of All the Afflicted. Only since 2010, however, has the Sunday Mass started to be celebrated regularly in Belarusian several times a year in Grodno. The same year, Saturday morning Divine Liturgy at Saints Peter and Paul Cathedral in Minsk became to be served in the Belarusian language, too. At this church, the sacrament of baptism, wedding, and funeral can also be led in Belarusian at parishioners' request. In 2013, Orthodox media reported that "for the first time in the history of Minsk the Sunday Liturgy was served in Belarusian".[12] Two years later, in March 2015, at the parish of Saint Archangel Michael, located on the outskirts of Minsk, the Liturgy was also served in Belarusian for the first time.

Such uses of Belarusian by the Orthodox Church can be viewed as the very beginning of penetration of Belarusian into Church activity. They may be a small concession won by the Belarusian-minded Orthodox community, but they do not alter the dominant status of Church Slavonic in the Orthodox Church. And while in national-minded media every new Orthodox parish at which religious service was conducted in Belarusian is presented as small "step" towards national Orthodoxy, Orthodox clergy explains that the scarce use of the Belarusian language in the Orthodox Church is justified not by its language policy, but by lack of interest among parishioners. Indeed, according to the results of the 2009 census in Belarus, only 21 per cent of the population acknowledged using Belarusian on a daily basis.[13]

Formally, the 1917–18 Local Council of the Russian Orthodox Church approved the use of languages other than Church Slavonic in liturgy in the Russian Orthodox Church. Moreover, in practice, several languages of small ethnic minorities – such as Finnish, Mordovian, Chuvash, Moldovan, Tatar, and Buryat – were used in liturgy in Russian Orthodox Church already in the nineteenth century. These languages were allowed in liturgy on the ground that people of these nationalities could not understand Church Slavonic and Russian. This argument obviously could not be applied to Belarusians, Ukrainians, or Russians, and, for them, Church Slavonic remained the only liturgical language.

In the new era of state independence, the language policy acquired new ideological dimensions. The return to the traditional concept of the Russian Orthodox Church's constituencies understood as the "eternal unity of the Russian, Ukrainian and Belarusian people" after the fall of Soviet Union was opposed by the ideas of sovereign nations developed by both Ukrainian and Belarusian national ideologies. In this context, Church Slavonic became a way of manifesting the institutional, canonical, and liturgical primacy of the Russian Orthodox Church, uniting these people under a common religious tradition.

Presently, several religious organizations represent national alternatives to the Belarusian Orthodox Church of Moscow patriarchate. One of them is Belarusian Orthodox Autocephalous Church, though it has no state registration. (Non–Moscow patriarchate Orthodox Christian communities can register with the Belarusian state only with the approval of a local Moscow patriarchate bishop,[14] because the Belarusian Orthodox Church of Moscow patriarchate registered its title as a brand name in 2001 so that no other organization can register using the name Orthodox Church in Belarus.) The Belarusian Autocephalous Orthodox Church (BAOC)[15] was founded in 1994. From the very beginning, it used only Belarusian and declared its mission the revival of a national version of the Orthodox tradition. It repeatedly tried to register itself in Belarus but was refused any legal status. Being practically banned in Belarus, this Church continues its underground activity, actively cooperating with the democratic political opposition and presenting itself as a religious organization, which "from the Christian position . . . defends the objective (meaning real) interests of the Belarusian people".[16]

According to some sources, under the jurisdiction of the BAOC, there were twelve parishes and religious groups by 2004. In the current political situation in Belarus, the BAOC is denied the opportunity to develop in the official public scene.[17] Belarusian media regularly report on the persecution of BAOC leaders for their religious actions. Overall, BAOC activity is characterised by open involvement in political opposition activity and driven by the idea of becoming the alternative religious organization which will unite all true Belarusian clergymen, will be supported by Belarusian society and, in spite of all difficulties, will become the "genuine national Church of Belarusians".[18]

Belarusisation policy in the Catholic Church

Catholicism is also an officially recognized part of the Belarusian historical legacy. According to the 2002 version of the Law on Freedom of Conscience and Religious Organisations,[19] several religious denominations were listed as particularly important for the historical development of Belarusian cultural tradition. First place in the hierarchy of traditional religions has been given to the Orthodox Church as having a determining role in the development of the traditions of the Belarusian people; the Catholic Church was mentioned in second place as playing a "spiritual, cultural and historical role on the territory of Belarus"; evangelical Lutheranism, Judaism, and Islam are listed as a group of faiths which are "indivisible" from the common history of the people of Belarus.

Catholics in Belarus enjoy the same rights as Orthodox believers; Christmas and Easter are state holidays, in accordance with both Orthodox and Catholic calendars. The apparent domination of Orthodoxy in post-communist Belarusian society as shown by statistics – 80 per cent of religious people are Orthodox – is, in practice, balanced out by the wide activity of the Catholic Church. Moreover, research has shown that in the western part of the country, where the majority of Catholics reside, religion is adhered to more strongly and more often plays a guiding role in social life (Danilov and Martinovich 2002: 7). For example, in Grodno province, 57.7 per cent of respondents declared to be religious and 12.5 per cent declared being non-believers. In the eastern part of Belarus, in Vitebsk province, the number of believers among Belarusians is 18.7 per cent, while 49.9 per cent are non-believers (Avsievich et al. 1999: 12). The difference can partly be explained by the fact that the communist ban on religion in the eastern part of the country stretched for seventy years, compared to forty in the western provinces.

The language policy of the Catholic Church, the second largest religious denomination in Belarus today, appears in stark contrast to the Orthodox Church. The official language of the Catholic Church, once viewed as an agent of expansion of "Polishness", is now unconditionally Belarusian. This policy was first formulated in the late 1980s. Previously, the presence of Catholicism served as a reminder of Polish domination over the territory of Belarus: it was the religion of the Poles, and the Polish language was almost unconditionally accepted as natural to Catholicism. The fact that there are also ethnic Belarusians among the Catholics was largely ignored by clergymen and Church authorities alike. In 1987 Uladyslau Charniauski wrote about Catholicism in Belarus: "Alien prayers, alien church, alien faith, alien priests, alien Mother of God and Lord God in alien, or at least, the Polish One" (cited in Yankouskaya 2007: 152).

Meanwhile, the Vatican had already declared its pro-Belarusian stance in 1950, when Radio Vatican started to broadcast in Belarusian. The Belarusisation of the Catholic Church in Belarus itself started in the late 1980s. Pope John Paul II had used the Belarusian language and demonstrated his support for Belarusian national Catholicism. In 1991, Belarusian was accepted as a working language in the Catholic Church in Belarus. Kazimir Świątek favoured the introduction of Belarusian into liturgy and prohibited the display of Polish national symbols in Belarus's Catholic churches. From the perspective of the Belarusisation of the Catholic Church, a problem remained with some clergymen of Polish origin, who did not separate the concepts "Catholicism" and "Polishness". They opposed the Belarusian language, seeing it as a means of "Belarusisation" of Poles living in Belarus. Today, antagonism against pro-Polish Catholicism comes from the local clergy and from believers. This milieu actively shapes the idea of Belarusian patriotism for Catholics and works towards the revival of national culture. One can observe a high level of patriotism among young people, with their intention 'to civilise Belarus and join Europe'. In their opinion, "the Orthodox Church, with its pro-Russianness . . . only prevents this" (Serdiuk 2002: 31). From this perspective, one of the most urgent needs discussed in the context of further Belarusiasation of the Catholic Church in Belarus became the establishment of a Catholic Theological Academy, which would ensure that Belarusian clergymen have been trained in Belarus and in Belarusian. Currently, Catholic priests in Belarus, if they are to pursue their education, need to go abroad, and the most popular destination in this case becomes Poland.

The idea of a Belarusian Catholic Theological Academy was first publicly voiced in 2008, after the meeting of President A. Lukashenka with Torcizio Bertone, the Secretary of State of His Holiness the Pope. At that meeting, the forthcoming talks on the concordat between Belarus and the Holy See were also announced.

The Catholic Church carries out a wide range of educational and publishing activities in Belarusian. All Catholic magazines and books are published in Belarusian (*Nasha Vera* (*Our Faith*), *Ave Maria*); basic religious texts, as well as prayers and songs, are translated into Belarusian. The Catholic clergy most often use the Belarusian language in their communications with parishioners. At the same time, most Catholic parishes both in the west and in the east of Belarus have held at least one Sunday Liturgy in Polish along with one or more Liturgies in Belarusian.

In the language policy of Catholic Church in Belarus, one can discern the perception of Belarus as only a small nation on a global map of Catholicism. This policy implies that the "stages" and "markers" of national identity that worked out in various Eastern European nation-building projects can be expected to start unfolding in Belarus as well. The Catholic Church in Belarus readily accepted its role of a facilitator in this process of national rebirth, and its language policy aimed at Belarusisation of Catholic community life provides clear evidence for this. This approach, however, did not take into consideration Belarusians' hybrid linguistic identity that has enabled the majority of Belarusians to avoid alienation and polarity in their perception of Russian language (Bekus 2014: 48). The Catholic Church's language policy has turned out to be more national-minded than that of the Belarusian society itself. At the same time, due to the century-long opposition between Catholicism and Orthodoxy in Belarusian lands, as well as cultural and civilizational association of Catholicism with Western tradition, Catholicism has little chance to influence Belarusian society outside its constituency.

The idea of the Uniate Church experienced a revival in the early 1990s, and was connected with the spiritual search of the Belarusian intelligentsia for a national church in which the liturgy and rituals would be united with the national symbols, traditions, and language. The first Belarusian Uniate parish was registered in 1992. The political scene saw the emergence of a number of pro-Uniate politicians and scholars: "Attempts to revive the Uniate Church at the present time should receive support among self-conscious Belarusians, because the Orthodox Church and Catholicism are practically not involved in the process of our national cultural resurgence" (Lych 1992: 68). This idea, however, had little chance to be realized in practice, as the "Uniates" comprise a small group. There were only thirteen Uniate parishes registered in Belarus in 2006; they all had few members and were often adjoined to a Catholic Church.

Conclusion

Religious revival in Belarus became not only a "tool for building identity in a changing world" (Borowik 2006: 272), but also a basis for building bridges with the historical past, restoring historical memory, and developing historical traditional heritage, both spiritual and practical (Titarenko 2008: 251–2). In the Belarusian context, however, the main Christian denominations – the Orthodox, Catholic, and Uniate Churches – are associated with different periods of history and different traditions. In the past, shifts in political authority and cultural domination were followed by upheavals in the religious history of Belarusian lands and, in some cases, resulted in entire reformatting of the religious identity of people.

Highlighting Orthodox tradition carries an appeal for Slavic unity, with reliance on Russian culture and language as an indivisible part of the Belarusian historical and cultural legacy. The acceptance of Orthodoxy as the genuine faith of the Belarusian people, constituting the spiritual core of the national tradition, provided additional arguments in favour of the pro-Russian geopolitical orientation chosen by the authorities. The Catholic Church, as well as the Uniate and Protestant ones, emphasizes the place of European heritage in Belarus's past. Their legacy in the Belarusian tradition accentuates the historical meaning of GDL and the Polish-Lithuanian Commonwealth, when the Belarusian tradition developed within the European framework.

In the wider European context, the split between Rome and Constantinople can be perceived "as an element that brings devotional diversity" to the region (Porter-Szűcs 2010: 17). In Belarus, however, this split variegates the religious contexts of Belarusian national unity. The Orthodox and Catholic Churches appeal to Belarusians' national self-determination and strive to provide a moral basis on which to build a Belarusian national identity. The language policies of these churches in Belarus demonstrate that they not only bring with them opposing geopolitical orientations, but also promote concepts of a national tradition that are differently seen as "truly Belarusian". In this way, contested religious framings of national identity enforce the role of religion as a basis for sub-national "groupness", which precludes it from being a means of national consolidation.

Notes

1 In post-Soviet conditions, religious identification, as some studies of post-Soviet religiosity reveal, is not limited to the dualism of "believers" and "atheists". There are also "quasi-religious groups" (believing in some supranatural forces) as well as those who hesitate in their religious identification (believing to some extent, but not considering themselves religious) (Rotman et al. 2009: 146, 151).

2 Polish-Lithuanian Commonwealth (1569–1795) – the state that was created after unification of the Kingdom of Poland and the Grand Duchy of Lithuania (GDL). The territory of GDL incorporated Lithuania, Belarus, and western Ukraine. GDL is often viewed as the predominantly Belarusian state. The Ruthenian population (believed to be predecessor of Belarusians) constituted the majority of the population of GDL, and Old Ruthenian was its official language.

3 Besides apostolic administrators for Moscow, ones for Novosibirsk and Karaganda were also appointed, yet the CRA deputy chairman, M. Ivolgin, refused to recognize the authority of the Moscow appointee, Tadeusz Kandrusiewicz, on the grounds that Moscow had not been consulted (Anderson 1994: 196).

4 The eastern region of Poland during the interwar period was populated mostly by Belarusians and Ukrainians. Part of this territory was correspondingly joined to the Ukrainian and Belarusian Soviet Socialist Republics in 1939.

5 The full text of the agreement is available on the BOC website: http://exarchate.by/resource/Dir0009/Dir0015/. Accessed 15 May 2018.

6 Ibid.

7 Official name: Brotherhood in Honor of Vilnia Martyrs Anthony, John and Eustaphy of the Belarusian Orthodox Church.

8 For the history of the Brotherhood, see http://sppsobor.by/bractva/vilna/409#more-409. Accessed 15 September 2014.

9 Between 1993 and 1995, the name of the periodical was *Pravaslaue u Belarusi i svecie* (*Orthodoxy in Belarus and around the World*).
10 *Pravaslaue* 1993, 1.
11 "U sabory praishlo abmerkavan'ne vykarystan'nya rodnai movy u Carkvie": http://sppsobor.by/life/news/news-sobor/10142
12 http://sobor.by/page/Vpervie_v_istorii_Minska_v_voskresniy_den_bila_otslugena_Liturgiya_na_belorusskom_yazike
13 National Statistical Committee of the Republic of Belarus. Accessed 15 September 2014. http://belstat.gov.by
14 "Belarus: Belarusian Orthodox Church". *Forum* 18, 6 November 2003. Accessed 15 September 2014 www.forum18.org/archive.php?article_id=177
15 At the beginning, it was named the Belarusian National Autocephalous Orthodox Church (Belaruskaia Narodnaia Autakefalnaia Pravaslaunaya Carkva). Accessed 15 September 2014 www.belapc.org/
16 "Golas Carkvy Pravaslaunai Autakefalii" 2006, 24. Accessed 8 September 2016. http://nashaziamlia.org/2006/05/29/114/
17 "Golas Carkvy Pravaslaunai Autakefalii" 2006, 24. http://nashaziamlia.org/2006/05/29/114/
18 "Da historyi Belaruskai Pravaslaunai Aukefalii", Belarus' – nasha ziamlia, June 2006. Accessed 7 September 2016. http://nashaziamlia.org/2006/05/29/114/
19 The full text of the agreement is available: http://pravo.levonevsky.org/bazaby/zakon/zakb1339.htm Accessed 24 April 2018.

References

Adamkovich, Ales'. Interview with Zianon Pazniak. "Zianon Pazniak: Naibolshuyu prykhilnasc' syarod moladzi maye katalicyzm" [Zianon Pazniak: Catholicism is the most attractive among youth], Tavarystva Belaruskai kultury u Litve. 8 June 2009. Accessed 6 April 2013. http://westki.info/tbk/7962/zianon-pazniak-naybolsuiu-prychilnasc-siarod-moladzi-maie-katalicyzm

Akintchyts, Stanislau. *Zalaty vek Belarusi* [Golden age of Belarus]. Khabarovsk: Novyi vzglyad, 2001.

Alekseyev, L.V. *Polotskaya zemlya (Ocherki istorii severnoy Belorusii) v IX–XIII vv.* [The Polatsak Land (an outline of the history of Northern Belarus) in the ninth–thirteenth centuries]. Moscow: Nauka, 1966.

Anderson, Benedict. 1983. *Imagined Communities.* London: Verso.

Anderson, John. *Religion, State and Politics in the Soviet Union and Successor States.* Cambridge: Cambridge University Press, 1994.

Arlou, Uladzimir and Saganovich, Henadz. *Dzesiats' viakou belaruskai historyi 862–1918.* [Ten centuries of Belarusian history 862–1918]. Vilnia: Nasha Buduchynia, 1999.

Avsievich, M., Zemlyakov, L., and Savostenok, L. *Religiya v usloviyakh socialnykh peremen v Belarusi* [Religion in the conditions of social changes in Belarus]. Minsk: Academiya poslediplomnogo obrazovaniya, 1999.

Bekus, Nelly. "'Hybrid' Linguistic Identity of Post-Soviet Belarus." *Journal on Ethnopolitics and Minority Issues in Europe* 13, no. 4 (2014): 26–51.

Bekus, Nelly. "On the Political Mission of Orthodoxy in Belarus and Its Consequences on the Church and State". In *Orthodoxy versus Post-Communism? Serbia, Belarus, Ukraine and the "Russkiy Mir"*, edited by Michał Wawrzonek, Nelly Bekus, and Mirella Korzeniewska-Wiszniewska, 70–158. New Castle: Cambridge Scholar Publisher, 2016.

Borowik, Irena. "Orthodoxy Confronting the Collapse of Communism in Post-Soviet Countries", *Social Compass* 53, no. 2 (2006): 267–78.

Chigrinov, P.G. 2004. *Ocherki istorii Belarusi*. Minsk: Vysheishaia shkola.

Danilov, A.V., and Martinovich, V.A. 2002. "Osobennosti konfessionalnoi struktury Respubliki Belarus" [Specifics of the denominational structure of the Republic of Belarus]. In *Poskommunisticheskaia Belarus' v protsesse religioznykh transformatsii* [Postcommunist Belarus in the process of religious transformation], edited by A.V. Danilov, 6–13. Minsk: Adukatsyia i vykhavanne.

Ermalovich, Mikola. *Pa sliadakh adnago mifa* [Tracing one myth]. Minsk: Navuka i tekhnika, 1991.

Halbwachs, Maurice. *Spoleczne ramy pamieci* [Social framework of memory]. Warszawa: PWN, 1969.

Hobsbawm, Eric. *The Age of Revolution 1789–1848*. New York: Mentor, 1962.

Hroch, Miroslav. 1985. *Social Preconditions of National Revival in Europe. A Comparative Analysis of the Social Composition of Patriotic Groups among the Smaller European Nations*. Cambridge: Cambridge University Press, 1985.

Ihnatouski, Usievalad. *Karotki narys historyi Belarusi* [Short outline of the history of Belarus]. Minsk: Belarus, 1991.

Kotliarov, I.V. *Respublika Belarus' v konfessionalnom izmerenii* [Republic of Belarus in denominational dimension]. Minsk: MIU, 2004.

Krutalevich, Vadim. *Rozhdenie Belorusskoi Sovetskoi Respubliki. Razvertyvanie natsionalno-gosudarstvennogo stroitelstva. Noiabr 1918 – fevral 1919* [The birth of the Belarusian Soviet Socialist Republic. Unfolding of national-state building. November 1918–February 1919]. Minsk: Nauka i tekhnika, 1979.

Lych, Leanid. "Religia i natsiianal'naia samasviadomast' Belarusau" [Religion and national self-awareness of Belarusians], *Belarusika Albaruthenica* 2 (1992): 58–69.

MacLaughlin, Jim. *Reimagining the Nation State: The Contested Terrains of Nation-Building*. London: Pluto Press, 2001.

Marples, David. *Belarus: A Denationalized Nation*. Amsterdam: Harwood Academic, 1999.

Martin, Terry. *The Affirmative Action Empire: Nations and Nationalisms in the Soviet Union 1923–1939*. Ithaca and London: Cornell University Press, 2001.

Mironowicz, Eugeniusz. "Kształtowanie się struktury narodowościowej w BSSR (1921–1939)" [Formation of the national structure of BSSR (1921–1939)]. *Bialoruskie Zeszyty Historyczne* 22 (2004): 42–57. Accessed 6 April 2013. http://kamunikat.org/usie_czasopisy.html?pubid=2196

Mironowicz, Eugeniusz. "Naciyanalnae pytanne u politycy pravaslaunai carkvy Zakhodnyai Belarusi u 1921–1939" [National issue in policy of Orthodox Church in Western Belarus in 1921–1939], In *Khryscianstva u histarychnym liose belaruskaga naroda* [Christianity in the historical destiny of Belarusian people], edited by S.V. Marozava et al., 316–20. Grodna: GrDU im Ya. Kupala, 2008.

Porter-Szűcs, Brian. "Christianity, Christians, and the Story of Modernity in Eastern Europe". In *Christianity and Modernity in Eastern Europe*, edited by Bruce Berglund and Brian Porter-Szűcs, 4–35. Budapest: Central European University Press, 2010.

Pospilowski, D.V. *Russkaya Pravoslavnaya Tserkov w XX veke* [Russian Orthodox Church in the twentieth century]. Moscow: Respublika, 1995.

Richters, K. *The Post-Soviet Russian Orthodox Church: Politics, Culture and Greater Russia*. London and New York: Routledge, 2012.

Rotman, D.G., Danilov, A.N., Bulynko, D.M., Haldman, L., et al. *Tsennostnyi mir sovremennogo cheloveka. Belarus v proekte "Issledovanie evropeiskikh tsennostei"* [Axiological world of the contemporary individual. Belarus in the project "The Study of the European Values"]. Minsk: BGU, 2009.

Salecl, Renata. "National Identity and Socialist Moral Majority". In *Becoming National*, edited by G. Eley and R. Suny, 418–25. New York, Oxford: Oxford University Press, 1996.

Serdiuk, V.V. "Rimo-katoliki Belarusi 1991–2001" [Roman Catholics of Belarus 1991–2001]. In *Poskommunisticheskaia Belarus' v protsesse religioznykh transformatsii* [Postcommunist Belarus in the process of religious transformation], edited by A.V. Danilov, 27–35. Minsk: Adukatsia i vykhavanne, 2002.

Smith, Anthony. *National Identity*. Harmondsworth: Penguin Books, 1991.

Snyder, Timothy. *The Reconstruction of Nations: Poland, Ukraine, Lithuania, Belarus, 1569–1999*. New Haven and London: Yale University Press, 2003.

Titarenko, Larissa. "On the Shifting Nature of Religion during the Ongoing Post-Communist Transformation in Russia, Belarus and Ukraine", *Social Compass* 55, no. 2 (2008): 237–54.

Tsvilik, M.B. "Sovremennaya konfessionalnaya zhizn' Belarusi" [Contemporary denominational life in Belarus]. In *Voprosy sovesti i religioznykh organizacii w Respublike Belarus'* [Issues of conscience and of the religious organizations in the Republic of Belarus], edited by M.B. Tsvilik and V.I. Novickii, 308–23. Minsk: "Chetyre Chetverti", 2005.

Vakar, Nicholas. *Belorussia: The Making of a Nation, a Case Study*. Cambridge, MA: Harvard University Press, 1956.

Veer, Peter wan der. *Religious Nationalism: Hindus and Muslims in India*. Berkeley: University of California Press, 1994.

Waring, A.G. "The Influence of Non-Linguistic Factors on the Rise and Fall of the Old Byelorussian Literary Language." *The Journal of Byelorussian Studies* 4, no. 3–4 (1980): 129–47.

Yankouskaya, V.V. "Mounae pytanne u Rymska-Katalickim kastsiole Belarusi" [Language issue in the Roman Catholic Church of Belarus]. In *Rossiiskie i slavianskie issledovania* [Russian and Slavonic studies], edited by A.P. Sal'kov et al., 152–62. Minsk: BGU, 2007.

Yazykovich, Larissa. "Religiinae zhytstsio belaruskai emigrathyi" [Religious life of Belarusian emigration], *Belarusika Albaruthenica* 2 (1992): 280–8.

Yekadumau, Andrei. "The Russian Factor in the Development of Belarusian Culture". In *Belarus-Russia Integration*, edited by V. Bulgakau, 169–220. Minsk and Warsaw: Analytical Group, 2003.

Zaprudnik, Jan. *Belarus at a Crossroads in History.* Boulder, CO: Westview Press, 1992.

10

THE PARADOX OF THE TERM 'DEMOCRACY' IN ARABIC DISCOURSE ON THE ISLAMIC MOVEMENTS

Kerith Miller

> The relationship between Islam and 'democracy' dominates much of contemporary Islamic political thought, particularly amongst Western-educated and 'modernist' Muslim intellectuals. However, much of the debate either does not bother to define 'democracy' or uses simplistic, naive definitions that people are undoubtedly accustomed to but which have little meaning or relevance if critically assessed. The problem is that the word 'democracy' has come to mean all things to all people; it has become so vague that it is virtually meaningless, even in the Western political culture from which it originated (Siddiqui 2003).

In the second half of the previous century, discussion in the Arab world on the topic of democracy increased at a rapid rate as the United States gained dominance in global affairs. This discourse has intensified in the present century, particularly with the concept of the stated mission of former US president George W. Bush to "democratize Iraq". Meanwhile, discourse on the development of the various branches of the Islamic movements has become progressively more sophisticated and interwoven with that of democracy as the movements themselves seek to establish goals, identity, legitimacy, and power within the political systems in the Arab world.

Methodology

This qualitative study of the use of *ad-dīmoqrāṭīya* in eight articles written for acclaimed newspapers and journals and published online, in addition to one YouTube video, examines the use of the term *ad-dīmoqrāṭīya* as it occurs in combination with its surrounding semiotically charged collocations and phrases. In each given instance, is *ad-dīmoqrāṭīya* positioned in an inclusive or an exclusive position towards the receivers? Which words are it – or adjectives describing it – placed in parallel structure with, thereby signalling the semantic intention of the author? What terms are subordinated to it, and to what terms is it subordinated within the sentence? In what manner is it connected to or located within other oppositions laid out in the discourse? The study draws out the semantic range of this word as it is manipulated in the context of the Islamic movements, highlighting contradictions both between polarities of the semantic range and within the discourse of a single person in order to create layers of persuasion against a background of ideological struggle.

The articles are selected on the basis of being at the forefront of a Google search in Arabic for the terms *ad-dīmoqrāṭīya* and *al-ḥarakāt al-ʾIslāmīya* ('the Islamic movements'). In the section 'Detailed analysis of a single article', one article is examined in detailed context for all uses of the term 'democracy' to provide a thorough example of the voices employed by one single author. In Section 3, the uses of this term are extracted from the other sources and compiled into categories of voice for analysis. They originate from various countries within the Arab world. The authors/speakers happen to be all male, with the exception of one female (Iqbal Al-Mu'min). The purpose of this study is to gather the information available online to a broad Arabic-speaking audience and examine the complexity of the voices found within this information on the topic of 'democracy', particularly in relation to the Islamic movements. The YouTube video of the episode on democracy of the popular Al-Jazeera programme *al-Ittijah al-Muʿakis* (*Opposing Views*) is selected because it is the single video with a strong focus on this topic found in an internet search at the time this study was carried out; November and December 2009.

Theory

The terms 'voice' and 'stance' can be employed almost interchangeably, and they require definitions specific to their use in this study. Both refer to the author or speaker's position in relation to the concepts discussed; however, 'stance' carries a stronger level of commitment, whereas 'voice' is examining a position and can range in tone from full support to sarcasm. Therefore, though in each of the articles examined[1] the author has a particular stance, each author speaks through a number of voices which are widely recognizable in the discourse on democracy in relation to the Islamic movements, even if the author intended to discredit a particular voice.

The voices are considered from the perspective of Bakhtin's theories on speech genre and polyphony. Bakhtin (1981: 294) pointed out that

> language is not a neutral medium that passes freely and easily into the private property of the speakers' intentions; it is populated – overpopulated with the intentions of others. Expropriating it, forcing it to submit to one's own intentions and accents is a difficult and complicated process.

Semiotically these concepts are considered in light of Judith Irvine and Susan Gal's (2000) theory of *iconization* and the manner in which the authors/speakers connect the concept 'democracy' to words with sign values selected to situate themselves on the side of common sense, presumably in line with the interests of the reader/listener. Charles Briggs and Daniel Hallin discussed the idea that "discourses create the publics that appear in those discourses as pre-existing" (2007: 46), a phenomenon which helps explain how authors use language indirectly in order to quietly convince readers that a particular opposition is a socially accepted fact.

Briggs and Hallin additionally cited Silverstein's use of the term 'misrecognition' to elucidate this process, explaining that

> European critical discourse analysis envisions discourse both in Foucauldian perspective, as practices that produce the objects to which they purport to refer, and through Bourdieu's practice theory, as infusing symbolic forms with value, citing their acquisition within particular institutions, controlling access to them, and naturalizing their categories and inequalities.
>
> (2007: 45–6)

Finally, these semiotic processes are viewed through the lens of Victor Turner's *self-categorization theory*, which clarifies the means by which authors create such categories of oppositions and treat them as natural. The aim of this study is to illustrate both in depth and in breadth a cross-section of this universally applied linguistic process as it occurs in this particular discourse on democracy.

Detailed analysis of a single article

One of the most essential aspects of the paradox involved in the use of the term 'democracy' is the desire of scholars writing about the Islamic movements to find within the boundaries of the concept the means to define a place of power within the region, allowing them to create or maintain public policies in accordance with their beliefs and often facing opposition from the ruling governments as well as other groups due to ideological and political differences. At the same time, it is crucial for those working in the name of Islam to maintain the supremacy of the Qur'ān and to avoid identifying too closely with values which are perceived to be contrary to Islam. Iqbal Siddiqui, former editor of *Crescent International*, pointed out the dangers as he perceived them in a June 2000 article:

> However, when Muslims talk in terms of 'democracy', even if they do not mean to be pro-western, the effect is a) to give the impression of endorsing western values and western claims to represent universal values; and b) to lay themselves open to accusations of being undemocratic by westerners. Then suddenly, instead of being able to simply accept the parts of 'democracy' that are common to Islamic political thought, and criticise the rest, Muslims find themselves having to explain why they do not accept all of democracy, at which they tend to be extremely apologetic and ineffective. They are put on the defensive and find themselves talking about democracy instead of Islam. The other problem is that they are then tempted to work alongside non-Muslims who claim to be democratic, on the basis that they have common agendas and principles. This also dilutes the Muslims' ability to put Islam first in their thoughts and actions.

Muslim writers who are generally in favour of the concept of democracy are required to take a double stance in order to present certain values, such as the peaceful attainment of power, and the right of people with diverse opinions to share the political stage, as helpful to their cause, and to turn away from the facets of the same values which could be perceived as threatening to Islam – for example, that these parties would have a voice but not absolute say. Writers who are generally not in favour of the concept of democracy and are sarcastic about it may still find it useful to hold as a standard for purposes of decrying 'undemocratic' behaviour. Dr Ahmad Jamil Azam, member of the Muslim Brotherhood in Jordan, proves himself to be a master of the double stance, which is the reason that his article, *Murabbaʿ al-ḥarakāt al-ʾIslamīya* ('The Square of the Islamic Movements') is selected for the thorough-analysis section of this study.

Contextual background

In *Murabbaʿ al-Ḥarakāt al-ʾIslamīya*, Azam divided present-day Islamic movements into four basic types (2006: 1): first, *al-ḥarakāt al-ʾIslamīya as-sīyasīya* ('the political Islamic movements'), which work with local community organizations and elections; second, *al-ḥarakāt at-takfīrīya* ('the expiatory movements'), which is a term used for the movements that, as Azam put it, "*tu'min wa tumāris al-'unf wa-l-'irhāb*" ('believe and practise violence and terrorism'); third, *ḥarakāt al-muqāwama al-'Islamīya* ('movements of Islamic resistance') that resist non-Islamic political and economic occupation of areas considered to be Islamic; and, fourth, *al-'Islam aṭ-ṭā'ifī* ('sectarian Islam'), which he considered the most recent branch with stronger ties to Shī'a than to Sunni Islam.

Azam's article defines three main goals (2006: 1). The first is the necessity to acknowledge the four categories, the second is the resolution to dismantle the square, and the third is the identification of the parties that need to be dismantled. The article traces the Islamic movements as they are currently known back to the beginning of the *Fataḥ* faction of the Palestinian Liberation Organization in the 1960s, claiming that the current quadrilateral form discussed is a recent, gradual development, and that even in the 1980s there was not a large distinction between schools of thought amongst the Islamic movements. In the 1980s, the two primary movements were *al-'Ikhwān al-Muslimūn* ('The Muslim Brotherhood') and *Ḥazb at-Taḥrīri-l-Islāmī* ('the Islamic Party of Liberation'), both of which patterned themselves after *Fataḥ*, with opposition to the occupation of Palestine a central motivation. It was later, in the 1980s, that the movements involving violence and terrorism appeared, such as those which could be found in Egypt (2006: 2), as distinguished from the opposition movements, for example Hazbullah in Lebanon – which were focused on securing rights and freedom for Palestinian people. Within this category, Hamas developed as the official wing of the Muslim Brotherhood in Palestine. By the beginning of the twenty-first century, Azam depicted the Islamic Movements as a triangle, involving the political side, the resistance side, and the side in favour of the use of violence against unbelievers. He claimed to see the current war in Iraq as the instigating factor of the fourth side, due to the central focus on sectarian issues.

Analysis

With this background, Azam explained why he believed it necessary to dismantle the 'square', and he introduced the term 'democracy':

> *ḥatā fī at-tīyār al-ghālib fī al-jamā'āti-l-'Islāmīyati-s-sīyāsīya, wa fī tīyārāt al-muqāwama al-islāmīya, hunāka rafḍ li-l'unf al-'ā'mā wa-t-takfīr, wa-l-istihdāfi-l-madanīyīn, wa hunāka taṭawwur fī-l-janāḥayn naḥū-t-ta'ātī ma'* ***al-mujtama'i-l-madanī wa niẓāmi-d-dīmoqrāṭī****. wa kadhalaka yatammu rafḍ al-'unfi-ṭ-ṭā'ifī wa-l-mumārisāti-ṭ-ṭā'ifīya min ḥaythu al-mabda'.*

> [even in mainstream thought in the political Islamic groups, and in the streams of thought in the Islamic resistance groups, there is refusal of blind violence and the charge of unbelief, and of the targeting of civilians, and there is development in the two sides towards dealing with **civil society and the democratic system**. In this way also, they refuse sectarian violence and the practice of sectarianism on principle.]
>
> (2006: 2)

Parallel structure

One prominent characteristic of the Arabic language is called ***parallel** structure*, or the use of two or more nouns or adjectives together that add to and specify the intended meaning within a sentence. In this case, then, Azam used 'democracy' as a distinctly favourable concept by placing it in parallel structure with 'civil society' and by placing it on the side opposed to the concepts of blind violence against unbelievers and the targeting of civilians. By qualifying the violence under discussion as blind, and parallel to 'targeting of civilians', Azam carefully positioned himself in support of the Islamic political and resistance movements and against sectarian violence, as well as those who advocate violence against unbelievers for being unbelievers. Still he concluded this paragraph with the idea that this undesirable level of violence is to some degree interwoven with all the Islamic movements, which is part of his reasoning for the call to dismantle those two sides of the 'square' (2006: 2): "*ʾjism al-ḥarakāt al-ʾIslāmīya, ḥatā fī* ***ʾaqṣā ḥālāti-l-ʾiʿtidāl*** *laysa barīʾan tārīkhīyan min wizrati-l-janāḥayn/a-ḍ-ḍilaʿayn al-ākhrayn, wa-l-ḥaraka laysat muḥaṣṣana ʾaw munfaṣila tamaman ʿanhuma*" [the body of the Islamic movements, even in **the most extreme cases of moderation**, is not historically innocent of the sin of the last two flanks/sides, and the movement is neither fortified against nor completely separated from them].

Though semantically the terms 'extreme' and 'moderate' imply distinctly different degrees on a continuum, Azam chose the word *ʾaqṣā* or 'furthest point, utmost degree, extreme' to describe within certain Islamic movements the high level of 'moderation', or willingness to work with people who may have opposing viewpoints. He contrasted this to such extremists who aim to avoid or annihilate those who disagree with them, though he cautioned that the latter (as exemplified by the expiatory and the sectarian groups) at times infect even the highest degree of the former.

Iconization

In setting up these graphic concepts, Azam participated in a process of linguistic ideology which Judith Irvine and Susan Gal (2000) labelled *iconization*. "Iconization involves a transformation of the sign relationship between linguistic features (or varieties) and the social images with which they are linked" (37). In his care to present the idea of democracy in the climate of cultural tension between 'the Middle East' and 'the West' in a manner that would be acceptable and even desirable to many of his readers, Azam always attached democracy to plurality and consistently opposed this to images of sweeping, needless violence ('blind', 'targeting civilians'). He avoided any direct connection of the term 'democracy' to Western culture per se, instead connecting it inextricably to the concept of 'civil society' and co-existence in a complexity of society in the Arab world.

Ideological oppositions

Azam drew a line between violence and opposition to occupation. He always placed 'violence' in combination with negative words like 'deviation' and 'justification' (below) and considered 'opposition' an important goal of the Islamic movements, in combination with goal-oriented words such as 'methods' and 'boundaries'. He summarized his distinctions specifically in his conclusion:

> *bi-kalimātin 'ukhra waẓīfati-l-'Islāmīyīn al-'ulā fī-l-marḥalati-l-muqbila hīya taṣaddur wa qīyādat juhūdi* ***taṣḥīḥi-l-'inḥirāfi-l-fikrī lidā ba'ḍi-l-'Islāmīyīn bi-tijāhi-t-takfīr wa-l'unf wa-ṭ-ṭā'ifīya****, wa murāja'a wa balwarat 'ususi-l-'alāqati-s-sīyāsīya ma' bāqī qiwā-l-mujtama'i-l-madanī, wa murāja'at ma'nā wa ḥudūd w-'asālīb 'amali-l-muqāwama fī-l-manāṭiqi-l-muḥtala, wa-t-tamyīz baynahā wa bayn al'irhāb, wa-t-tamyīz bayna-t-taktīkī-s-sīyāsī al-wāqa'ī, wa-sh-shi'ārāti-t-taḥshīdīyati-t-ta'bīīyati-l-'āma.*
>
> [In other words, the duty of the first Islamists in the next stage is the issuing and guidance of efforts to **correct the ideological deviation of several Islamists towards the justification of violence against unbelievers and sectarianism**, and the review and crystallization of the foundations of political connections with the rest of the forces of civil society, and review of the meaning and the boundaries and the methods of resistance in the occupied

> areas, and distinguishing between that and terrorism, as well as distinguishing between actual political tactics, and the general accumulated tactical slogans].
>
> (2006: 3)

Understanding Azam's main focus provides a clearer framework for his use of the term 'democracy' further in the article. He carefully addressed the issue of plurality which was at the crux of the paradox of the concept of democracy in conjunction with strict religious legal codes (2006: 2): "*wa lā buda min ḥasm qaḍāyāti-t-a'adudīya,* ***wa taḥdīdan at-ta'adudīya ma' al-jamā'āti-latī tatabanā fikran ghayr Islāmī,*** *wa* ***qabūl bi-ltizām bi-n-niẓām ad-dīmoqrāṭī*** *ḥatā law kāna-l-islāmīyūn 'aqalīyatan fīhi*" [And it is necessary to resolve the issues of plurality, and **specifically the plurality with the societies that adopt non-Islamic ideas**, and the acceptance of **commitment to the system of democracy** even if the Islamists should form the minority].

Culturally charged words

In his article, Azam turned away from distinctions between terrorism and terrorists, implying that there is a fundamental unity between ideology and practice and claiming that this must be recognized in order to really eradicate the kinds of violence he was opposed to. He was also insistent on disconnecting those who were proponents of this type of violence from Islam itself by careful manipulation of highly charged words such as *takfīr* and *fitna*:

> *lā zāla hunāka dākhil jamā'ati-l-'Ikhwāni-l-Muslimīn,* ***man yarfuḍu-t-ta'dudīyata-d-dīmoqrāṭīya wa-l-fikrīya wa-l-aqā'idīya,*** *wa lā yuqaddimu mawqifan wāḍiḥan . . . fa-tada'ī rafḍ at-takfīr wa-t-ta'āṭuf ma' at-takfīrīyīn, bi-dharī'ati mumārisāti-l-muqāwamāt, rughmi jarā'im ha'ulā' ḍid al-madanīyīn min Muslimīn wa ghayrahim, wa rughmi-****d-da'wa-ṣ-ṣarīḥa li-l-fitnati-ṭ-ṭā'ifīya****, min-ma yuthīru makhāwifa mubararatan ḥawl ṣidq iltizām al-Ḥarakāti -l-Islāmīya bi-l-lā'unf, wa-t-ta'dudīya, 'al-ān wa mustaqbalan, wa-fī 'athnā' wujūdihim khārij as-sulṭa 'aw dākhilihā.*
>
> (2006: 2)

> [There are still **those who reject plurality – democratic, intellectual, and ideological** – amongst the Muslim Brotherhood,

> and who don't present a clear position . . . and thereby claim to reject *takfīr* while sympathizing with the *takfīrīyīn*, under pretext of the practice of resistance, despite the crimes of these people against civilians both Muslim and non-Muslim, and despite **the explicit invitation to the seduction of sectarianism**, which arouses justified fears about the sincerity of the commitment of the Islamic movement to non-violence and plurality, now and in the future, whether outside authorized power or within it.]

Azam maintained that it was not enough to claim one was against the practice of violence in the manner which he labelled *takfīr*, as well as sectarian violence; one must also guard against those who condone it and practise it in the name of policy or resistance. Again he was connecting democracy with plurality, and in this sentence in fact subordinating it by including it as one of three modifiers – 'democratic', 'intellectual', and 'ideological'.

The use of the words *takfīr* and *fitna* in such a context also helps anchor the reader into Azam's stance on the issues. *Takfīr* carries both the meaning 'atonement' or 'expiation' and the meaning 'seduction to infidelity' and 'charge of disbelief'. Clearly Azam used it to express a concept that, though related, goes beyond these definitions, because he always paired *takfīr* with violence, whether in parallel structures, in his internal definitions, or in a broad context. He consistently opposed it to the idea of resistance at a level of non-violence or a level that did not attack civilians. He employed this negative connotation of *takfīr* and *fitna* to lend power to his call for dismantling the square:

> *ʾammā-l-jihatu-l-masʾūla ʿan tafkīki-l-murabaʿ, fa-lā buda min at-tanwīh ʿanna-t-tafkīk yaʿnī ʾawlan, barāʾata-t-tīyārati-l-ʾIslāmīyati-s-sīyāsīyati-l-madanīyati-t-taʿadudīya, wa-l-muqāwamati -l-mashrūʿiya, min al-ʾaḍlāʿi-l-ʾukhra, wa rafḍa-l-ʾindimāj ʾaw at-taqārub maʿhā fi-ʾiṭār wāḥid, wa rafḍ ʾafkārihā wa rafḍa-l-qāʾimīn ʿalayhā, ʾidhā mā waṣala-l-ʾamru bi-him* ***li-l'inḥirāf wa-l-jarīma wa-l-fitna b-ismi-d-dīn****. wa thānīyān, ʾistikmāla-l-murājaʿā-l-fikrīyati-lati badaʾathā-l-jamāʿa mundhu as-sabaʿīnīyāt,* ***bi-tijāh takrīsi-l-iltizām*** *bi-d-dīmoqrāṭīyati-t-taʿdudīya . . .*

> [However, from the aspect of the responsibility of dismantling the square, it is necessary to mention that dismantling means first, the exculpation of the political, civilian, and pluralistic Islamic

> currents, and the resistance project, from the other sides, and rejection of the coalition or convergence with them from one side, and refusal of their ideas and rejection of their leaders, if they don't want the situation to become **perversion and crime and sedition in the name of religion**. Secondly, it means completion of the intellectual examination that the community began in the seventies, **in the direction of the dedication to commitment to pluralistic democracy . . .]**
>
> (2006: 2)

Here Azam used the word *fitna* to connect the groups who were practising violence to the qualities which they were trying to practise it against – perversion and crime and sedition all being outside the realm of Islamic values. Then he returned to the word 'democracy' for the final time in his article, giving it for the first time the firm identity of a noun. He continued to modify it with the adjective 'pluralistic' and placed his strongest call for the necessity of these qualities within the boundaries of Islamic society.

Cross-analysis of multiple sources

Following the example of the uses of democracy as they occur in a full article is the broader examination of a cross-section of seven additional articles and a YouTube video, all written/produced within the last ten years. Within these sources, I have identified nine general voices and categorized the most salient examples of the speech token 'democracy', including as much context as necessary for the reader to recognize the stance of the particular voice.

Bakhtin, well known for his analysis of competing voices from a dialogic point of view, spoke of the concept of argument as "the layering of meaning on meaning, voice upon voice, strengthening through merging (but not identification) the combination of many voices (or a corridor of voices)" (1986: 121). The data compiled for this study represent exactly that – a corridor of voices, each voice weaving chains of signs and symbols into desired shapes, colouring and re-colouring the theme word, 'democracy', with qualifiers and placement in evaluative hierarchies. Though an entirely comprehensive qualitative analysis of every voice on this topic in the Arab world would be inconceivable, this slice represents the articles and video that emerged as the most prominent and widely viewed on the subject according to the popular Google search engine in November and December 2009. Therefore the voices are a highly representative sample of public educated discourse on the issue in the years leading up to the Arab Spring revolts.

List of sources used

Below is a numerical list of the sources used in this study, followed by their appearance in the aforementioned nine categories of voice:

(1) Ghalioun, Burhan: 2003. *ad-dīmoqrāṭīya al-'Arabiya wa bu'bu 'i-l-ḥarakāti-l-'Islāmīya*

[Arab democracy and the bogeyman of the Islamic movements]

(2) Abdul Jawad, Jamal: 2003. *al-ḥarakātu-l-'Islāmīyati jādatun fī tabannī-d-dīmoqrāṭīya wa lākinaha tafshalu wa tatahāwanu fī taṭbīqihā ḥāl wusūlihā li-s-sulṭa*

[The Islamic Movements are serious about building democracy, however they fail and compromise in its application in regard to acquisition of power]

(3) Al-Jazeera: 2004a. *al-'aḥzāb as-sīyāsīya fī-l-waṭani-l-'Arabī*

[The political parties in the Arab homeland]

(4) Al-Jazeera: 2004b. *al-ḥarakāt al-'Islāmīya wa-d-dīmoqrāṭīya: al-muwāqif wa-l-mukhāwifu-l-mutabādila*

[The Islamic movements and democracy: mutual attitudes and concerns][2]

(5) Abu Haniya: 2006. *al-ḥarakāt al-'Islāmīya wa-t-taḥāwulu-d-dīmoqrāṭīya*

[The Islamic movements and the democratic transformation]

(6) Abdul Hafiz, al-Mustafa. 2006. *al-harakāt al-'Islāmīya wa-l-'amalīyatu-d-dīmoqrāṭīya fī-l-'ālami-l-'Arabī: 'istikhshāfu-l-manātiqi-r-ramādīya*

[The Islamic Movements and the process of democracy in the Arab world – discovering the grey areas]

(7) Al-Mu'min, Iqbal. 2009. *ayahuma 'akthar ṭā'ifīyatan Ādam 'am 'Iblīs?*

[Which of the two is more sectarian, Adam or Satan?]

(8) *al-Ittijah al-Mu'ākis: ad-dīmoqrāṭīya* [*Opposing Views: Democracy*]

(Al-Jazeera programme, 29 August 2007)

Data

A. Aligned with the ideas that the moderate Islamic movements desire democracy; that this is valuable and achievable and can be compatible with Islam

ay **infitāḥ dīmoqrāṭī** jiddī sa-yakūn li-ṣāliḥi-l-ḥarakāti -l-ʾIslāmīya. (from source 1)

[Any serious **democratic emergence** will be for the welfare of the Islamic movements.]

ad-dīmoqrāṭīya hīya al-khīyār al-waḥīd aladhī yuqaddimu-l-furuṣa-ḍ-ḍurūrīyata-l-muḥtamala li-nazaʿi fatīli-l-ʿunf. (from source 1)

[**democracy is the single choice** that provides the most likely necessary opportunities to remove the fuse of violence.]

ʾahamīyatuhu al-qaṣwā laysa faqaṭ fī taṭawwur ʾafkāri-l-ḥarakāti -l-ʿIslāmīya, wa ʾinnama ʾayḍan fi **taṭawwuri-d-dīmoqrāṭīyatī nafsihā.** (from source 2)

[the highest importance lies not only in the development of the ideas of the Islamic movements, but also in **the development of democracy itself.**]

yuḥāwalu Nawhīḍ **nazaʿa-l-ʿaqdīya ʿan ad-dīmoqrāṭīyatī**-latī hīya fī-n-nihāya kamā yaqūl "mujarrad ālatin lā wāʿīya li-tanẓīm khilāfāti-n-nāsi salmīyan wa tartīb ikhtilāf wijhāt naẓarihim wa maṣāliḥihim fī sīyāq lā ʿunfī, fa-hīyā laysat dawlatan wa lā niẓām ḥukmi wa lā tashrīʿan wa lā tamuttu bi-ṣilatin li-l-ʿalmānīya wa-l-lībirālīya". (from source 4)

[Nahwid tried **to remove the aura of doctrine from democracy** that is in the end, as he says, "simply an instrument without consciousness; to regulate controversies of the people peacefully, and settle differences of opinion and interests in a non-violent context, for it is not a nation nor a regime nor legislation, and it has no relation to secularism and liberalism."]

B. Questioning the compatibility of Islamic versus democratic ideologies – ambiguity

hal hunāka khaṭar ʾIslāmī ḥaqīqī yuhadidu ay **taḥawwul fī-**t-tijāhi-d-dīmoqrāṭīya ʾaw hal yakūn al-ʾintiqāl naḥu niẓām ʾIslāmī yaqfal bāba-d-dīmoqrāṭīya . . .? (from source 1)

[is there genuine Islamic danger that threatens any transformation in the direction of democracy, or would a movement towards an Islamic system lock the door on democracy . . .?]

'inna ad-dīmoqrāṭīya 'aḥad 'ashkāli-l-ḥukumati-l-adhī tatawaffiru fīhī jumlatun min al-maʿāyīr: k-al-munāfisata-l-ḥurrata-latī tutīḥ at-tadāwula-s-salmī li-s-sulṭa, wa-l-mushārikata-s-sīyāsīya wa ḥaqqa-l-intikhāb wa-l-iqtirāʿ, wa ḍamāna-l-ḥuqūq al-insānīya wa-l-madanīya mithl ḥurrīyati-t-taʿbīr wa-ṣ-ṣaḥāfa

wa-l-inḍimām 'ila-l-'aḥzāb as-sīyāsīya wa takwīnihā, wa-masā'ilati- l-ḥikām wa-ḍamān ḥukm al-qānūn, wa-yumkin al-jizm bi-'an hādhihi-l-maʿāyīr lā tanṭabiq bi-ṣūratin kāmalatin ʿalā-l-ʿalami-l-ʿarbī fa-t-tahāwul ad-dīmoqrāṭīya ʿamalīyatun intiqālīyatun . . . (from source 5)

[In fact, **democracy is one form of government which offers a number of standards**: such as free competition that enables the peaceful transition to power, political participation, and electoral rights; secures human and civil rights like freedom of expression, freedom of the press, and the right to assemble and join political parties, as well as governmental issues such as the guarantee of government by law; and it is possible to assert that **these standards aren't completely in conformity with the Arabic world, as the democratic transformation is a transitional process** . . .']

muʿẓamu-l-ḥarakāti -l-'Islāmīyati-l-mutaʿdila lam tawali dabarahā tamāman l-'īdilojīyātihim al-'ūlā fa-bi-dākhil kūli-l-munaẓamāt lā yazāl hunāka mashāddat bayna-l-ahdāfi-l-qadīma dhāt aṣ-ṣila bi-inshā' dūl 'Islāmīya wa tatbīqi-sh-sharīʿati-l-'Islāmīya, wa bayna-l-ahdāfi-l-jadīda wa-l-latī minhā 'an **tuṣbiḥu tilka-l-ḥarakāt bi-mithābiti 'aṭrāf mu'athiratan fī 'iṭāri niẓām dīmoqrāṭī taʿdudī** wa-n-natījati-l-ḥatmīya li-tilka-l-mushāddāt hīya itisāʿ niṭāqi-l-ghamūḍi fī ʿadadin min al-qaḍāyā (al-mināṭiq ar-ramādīya). (from source 6)

[Most of the moderate Islamic movements have not turned their backs completely on their first ideologies, so within all organizations there are still altercations between the old goals, with a connection to building Islamic nations and the practice of Islamic law, and the new goals, amongst which is for the movements to become **influential parties in the framework of a pluralistic democratic system**, and the inevitable result of these altercations is a widening in the scale of ambiguity in several of the issues (the grey areas).]

C. Questioning the compatibility of Islamic versus democratic ideologies – stark opposition

'amma idhā istiṭāʿat al-ḥarakātu-l-ʿIslāmīya 'an tafraḍa haymanīyatahā bi-ghayri munāfisa ʿalā al-wāqaʿi-s-sīyāsī, fa 'innahā

lā yumkin bi ʾay ḥāl ʾan tanḥāz li-d-dīmoqrāṭīya bi-ghaḍi-n-naẓar ʿan al-qanāʿāti-l-fikrīya wa-l-fatāwā wa-t-tafsīrāti-l-fiqhīya. (from source 2)

[However, if the Islamic movements were able to impose their suzerainty on the political situation without competition, there is **no possibility in any circumstance of siding with democracy**, regardless of the ideological convictions and the fatwas and the religious interpretations.]

D. Critique of Arab governments for repressing democracy amongst Islamic movements by instigating fear of extremism from the movements entering their power system (defense by the governments of the 'old guard')

fī-ṣ-ṣirāʿi-l-ʿanīfi-l-jārī fī-l-bilādi-l-ʿarabīya ʾizāʾ ʾiṣlāḥ nuẓumi-s-sīyāsīya aṣbaḥa-t-takhwīf min al-Ḥarakāti -l-ʾIslāmīya wa iḥtimāl sayṭratihā ʻalā-l-muʾassasāti-t-tashrīʿīya yustakhdam ḥujatan raʾīsīyatan **li-tabrīr qamʾi-l-ḥarakāti -d-dīmoqrāṭīya wa-l-hifāẓ ʻalā-l-waḍaʿi-l-qāʾim.** (from source 1)

[In the current violent struggle in the Arab countries towards reform of the political systems, the threat of the Islamic movements and the probability of their dominance over the law-making organizations is used as an official excuse **to justify repression of the democratic movements and the preservation of the established situation**.]

wa rubbamā yakūn aṣ-ṣirāʿ al-jadīd al-adhī yadawaru bayna-l-ʾIslāmīyīn wa-n-nukhab al-**ḥākima ʾaḥad ʾahami-l-ʻawāʾiq alatī taqifu fī ṭarīqi-t-taḥāwuli-d-dīmoqrāṭī fī al-ʻalami-l-ʻArabī**, wa ma lam yaʿud aṭ-ṭarafān an-naẓar fī mawāqifihimā-l-mutaṣalliba fa-ʾinna-l-khasara sawfa tuṭāl al-jamīʻa, fa-l-āmāl ash-shʿabīya-l-ʻarabīya tanʿaqidu ʿalā ḍarūrati wujūd qīyādāt muʿatadala min kilā-l-muʿaskarīn mustaʻada li-l-inkhirāṭ fī ḥiwār shāmil yataraffaʻu ʻan al-mukāsib wa-l-muṣāliḥi-sh-shakhsīya wa-l-fiʾawīya-ḍ-ḍīqa. (from source 5)

[And perhaps the new struggle which is occurring between the Islamists and the ruling elite is **one of the most important obstacles which stands in the way of the democratic transformation in the Arabic world**, and in which both sides no longer examine their unyielding positions, and so the loss will extend to all, as the hopes of the Arab populace are interwoven with the necessity for the presence of just leadership from both camps prepared to join in comprehensive dialogue above the level of petty personal and party profits and interests.]

E. Viewing democracy as desirable but skepticism or sarcasm regarding the Islamic movements or the Arab countries themselves

al-'i'tiqād bi'anna ash-sha'ūba-l-'Arabīya laysat nāḍijatan ba'd
li-d-dukhūl fī-l-**manẓūmati-d-dīmoqrāṭīya.** (from source 1)
[the consideration that the Arab people are not yet mature enough to enter the **democratic system**]

wa-tabdū ḥālatu-l-waṭanu-l-'Arabī wafq hādha-l-mu'shir khālīyatan min at-tanāfusi-s-salmī 'alā-s-sulṭa, bal **yakād al-waṭanu-l-'Arabī yakūn istithnā'an min al-ḥālati-d-dīmoqrāṭīyatī**-latī habbat rīyāḥuha 'alā-l-'ālam fī-s-sanawāti-l-'ashra-l-'akhīra min al-Filibīn ilā 'Amrīkā-l-Lātīnīya wa min al-'Ittiḥādi-s-Sofīayatī ilā Janūb 'Afrīqīyā wa min 'Asīyā-l-ūsṭā ilā 'Awrobā-sh-sharqīya dūna 'an taḥarruk hādhi-r-rīyāḥ 'ay waraqa bi-l-manāṭaqi-l-'Arabīya fa-taẓhuru munaẓimatu-sh-Sharqi-l-'Awsaṭ ka'innahā khārij al-'ālam. (from source 3)
[And it seems as if the Arab nationalist position according to this indicator[3] are devoid of peaceful competition for power; rather, it seems that the **Arab homeland is almost an exception to the situation of democracy** that blew its winds on the world in the last ten years, from the Philippines to Latin America, and from the Soviet Union to South Africa, and from Central Asia to Eastern Europe, without these winds moving even a single paper in the Arab lands, thus creating the appearance that the region of the Middle East lies outside the world.]

wa nada'ī naḥnu fī balad dīmoqrāṭī ḥurr. fīyā li-naksati-d-dīmoqrāṭīyati-l-muzmī'a idhan! (from source 7)
[And we claim we are in a free democratic country (Iraq). Faithful to the setback of the planned democracy, then!]

F. Critique of ruling powers in Arab and Western countries alike

wa-l-wāqi' 'annahu laysat fi'āt al-musāliḥ al-ḥākima hīya waḥdihā-latī tastakhdam hādhihi-l-ḥuja li-tabrīri-l-ḥifāẓ 'alā-l-'amri-l-wāqi' wa-'iḍfā'u-sh-shari'a 'alā-l-istibdād wa-law bi-ṭarīqa salbīya, wa-'innamā 'ayḍan ad-duā'iru-l-'alīya wa makhaṭaṭū-s-strātijīyāti-l-kubrā fī-d-dūali-ṣ-ṣinā'a. fa-hum 'ayḍan ya'ataqiduna – raghmi-l-ḥadīth al-makrūr wa-d-dā'im mundhu fitratin 'an **mashārī'i dīmoqraṭati-l-munaṭiqi-l-'Arabīya** – bi'anna-ḍ-ḍaghṭa-sh-shadīd 'alā-l-'anẓimati-l-istibdādīya min 'ajil al-'iṣlāḥ yumkin 'an yaza'za' istiqrāriha wa-yaftaḥ aṭ-ṭarīq 'amām istīlā'i-l-ḥarikāti-l-'uṣūlīyati-l-mutaṭarifa wa-l-mu'ādīya bi-ṣarāḥa li-l-Gharb wa-li-l-muṣaliḥi-l-Gharbīya 'alā maqālīdi-sulṭa fīhā. (from source 1)

[And the reality is that it is not the band of ruling authorities alone that uses this excuse to justify the preservation of the existing situation and bestow legality on autocracy even if by negative means, but in fact also the high government agencies and the big strategic designs of the industrial nations. They also consider – despite the repetitive and perpetual conversation since some time ago on the **projects of the democratization of the Arab regions** – that the enormous pressure on the autocratic organizations in regard to reforms could potentially upset their stability and open the way for the appropriation of the reins of power by the fundamentalist extremist movements which are patently inimical to the West and Western interests.]

'a lam **yataṣāruf** Jorj **Bush wa** Toni **Blair ka ay mustabidin 'Arabī**? 'a lam yanfaridā fī-t-tikhadhi-l-qarārati-l-kubra dūna-l-'awda ilā ay mu'assasa? (from source 8)

[And don't **George Bush and Tony Blair behave like any Arab tyrant**? And don't they stand apart by making big decisions without consulting any organization?]

G. Skepticism regarding the idea of democracy – undesirable because it is a cover for Western hegemony

'a lam tataḥāwal '**Amrīka ilā 'a'tā diktatorīyatin** fī-t-tārīkh dākhilīyan wa kharijīyan? (from source 8)

[And isn't **America transforming into the most arrogant dictator** in history in the domestic and foreign spheres?]

H. Skepticism regarding democracy as it is practised in the West

'a laysa hūwa-l-mas'ūl 'an tarāja'i-d-dīmoqrāṭīya fī-l-Wilāyāti-**l-Mutaḥada** nafsiha? (from source 1)

[And isn't that what is responsible for the **decline of democracy in the United States** itself?][4]

'a lam ya'tarif Robin Kuk, wazīru-l-kharijī-l-Britānī-r-rāḥil
bi-'anna-s-sulṭa-t-tanfīdhīya taghawwalat kathīran wa-ṣaḥabat
al-bisaṭa min taḥt 'aqdāmi-l-barlamān? (from source 8)

[And didn't Robin Cook, former British foreign secretary, confess that the **executive power encroached a lot** and pulled the rug from under the feet of the parliament?]

'a lā ta'īsh Amrīka wa-ghayrihā taḥt **waṭa'i-l-qawānīni-ṭ-ṭuwāri'i?** (from source 8)

[And don't America and the others live under the **oppression of the emergency laws?**]

I. Questioning democracy as an idealism which is supposed to hold multiple idealisms together

iḍṭirār al-ḥarakāti -l-ʿIslāmīya li-t-taʿāyīsh fī bīʿa taʿdudīya ḥaqīqīya **sa-yajbaruhā ʿalā-l-qabūl bi-d-dīmoqrāṭīya.** (from source 2)

[The compulsion of the Islamic movements to co-exist in an essentially pluralistic environment **will force them to accept democracy.**]

Analysis

According to Bakhtin, "words belong to nobody, and in themselves they evaluate nothing. But they can serve any speaker and be used for the most varied and directly contradictory evaluations on the part of the speakers" (1986: 85). In article (1), for example, as we can observe in Table 10.1, Ghalioun (source 1) managed to represent the voice of six of these categories, skilfully winding point and counterpoint together to address the concerns of a multitude of readers and paint a more balanced view of the overall picture. How did he use the word 'democracy' to present these varying stances?

Briggs and Hallin discuss the pragmatics of social discourse as an ongoing process, "rendering some dimensions visible and construing them in

Table 10.1 Illustration of the cross-section of viewpoints

A	*B*	*C*	*D*	*E*	*F*	*G*	*H*	*I*
1	1	2	1	1	1	8	1	2
2	5		5	3	8		8	
4	6		7					

particular ways, erasing others, and imagining subjectivities, social relations, and forms of agency" (2007: 46). Ghalioun took a stance apart from the government and the Islamic movements, metapragmatically pointing to the discourse of governments who, by creating the 'bugaboo' of the aims of extremist Islamic movements and attempting to erase the voice of all Islamic movements in governmental procedures, in fact embodied a form of exactly what they were trying to scare the people with – dictatorial oppression.

Looking across all the voices represented here, one can see general patterns emerge from the categorized discourse segments – 'the moderate Islamic movements' are generally held to be connected with a desire for democracy, for example, but whether this in itself is associated with non-violence and human rights, or Western hegemony, or insincerity due to a hidden agenda to seize dictatorial power in the end depends on the stance of the author. Plurality is positioned as positive with democracy through connection with concepts such as non-violence, human rights, and peaceful ascension to power, but as negative if linked to domination of non-Islamic values and law in Islamic countries.

The complexity of the discourse is evidenced by the various dividing lines between in-groups and out-groups. Lütfiye Oktar discusses the importance of Turner's 'self-categorization theory' in understanding ideology in discourse across conflicting groups (2001: 315). According to this theory, by nature we as humans choose to represent our own group in the most favourable light, due to our need for self-esteem. We also draw oppositions in order to better distinguish ourselves from the 'other'. We then proceed to use language to create semiotic chains that fundamentally connect to icons of that which is 'good' and 'bad'.

> This implies that the representations of social arrangements, that is, the things that are considered as 'better' and the things that others stand for are associated with these polarized presentations of *us* and *them*. These social arrangements are descriptions of more general values. For instance, if 'secularism' is the general socio-cultural value, then 'freedom of religion and conscience' is one of the values a secularist ideology will present as something *we* stand for.
>
> (2001: 319)

The voices in the data demonstrate such oppositions, but looking across the stances that are represented by these voices, most are remarkably similar. In fact, it appears that a stance is missing – that of the extremist extremist public who would themselves claim that democracy is antithetical to Islam.[5] The closest to this voice would be Iqbal Siddiqui's quote (see Section 2) claiming that in the dialogue on democracy with non-Muslims, Muslims potentially stand to compromise their values.[6] However, the most common underlying stance is the presentation of the authorial voice of common sense that should align with the reader as a member of a populace whose interests (peace, justice, religious identity, a right to expression, etc.) could be in danger because of extremist religious groups, because of some ruling powers, or both.

Conclusion

In this light, it makes sense that underlying the polarities involved in the discourse on 'democracy' in relation to the Islamic movements is a paradox faced by all cultures that have attempted to work with the concept of rule of the populace. Simply because a majority wants something, does that make it right, or the best policy? Democracy is often equated with freedom in many languages, but freedom to *do* what, freedom *from* what, and freedom for *whom*? What if the majority of a community votes to use violence to solve problems? What are the necessary conditions for people with opposing ideologies to work together in a world of limited resources?

Sidorkin (2002: 70) pointed out Bakhtin's argument that

> [p]olyphony does not imply that 'the truth is somewhere in the middle'. . . . The very point of plurality for Bakhtin is constant touching, shifting, penetrating, mutual inclusion of voice. His idea that multiple voices never merge, never come to a grand ending of a grand narrative, doesn't in any way mean that the voices do not change each other. Quite the contrary, the interaction is the truest moment of their being.

Throughout this discourse, language (if not violence) is the tool by which people are working to convince others of the validity of their ideologies. When involved in a discourse of conflict, then, it is useful to examine the voices that appear in multiple guises from opposing stances and to look for the values which are held in common and relied upon to anchor the highly charged, elastic words such as 'democracy' to the desired position of the author. In this manner, it is possible to untangle and simplify webs of arguments and, ideally, search for satisfactory solutions to social issues.

Notes

1 Except in the instance of *al-Ittijah al-Mu'akis* (*Opposing Views*), in which the host, Faisal al-Qasim, presents a series of voices in the form of questions

representing one stance in a debate, followed by a series of voices representing the opposite stance.

2 Book review for book of same title published in 2000, edited by Ali al-Kawari, published in Kuwait by the publishing house *Dar al-Qirṭas.*

3 Islamic parties as an indicator of political diversity in the Arab homeland.

4 The increase of security measures in the United States due to terrorist attacks by extremist movements.

5 The one example under voice C, which would particularly express this stance, is written as a threat as to what would happen if religious fundamentalists were to attain political power, rather than a voice reflecting or supporting the stance.

6 The fact that a more extreme stance is missing from this data certainly does not mean that it doesn't exist; it simply alludes to the fact that in this group of widely viewed articles, the authors themselves consistently maintained a more moderate stance and appealed to values related to cooperation. It may also allude to the fact that an article defending a more polarized position in a discourse on politics will show up more easily in a search by sympathetic readers, according to the algorithms of the internet.

References

Abdul Hafiz, al-Mustafa. *al-ḥarakātu-l-'Islāmīya wa-l-'amlīyat ad-dīmoqrāṭīya fī-l-'ālami-l-'Araby: 'istikhshāfu-l-manātiqi r-ramādīya* [The Islamic movements and the process of democracy in the Arab world – discovering the grey areas]. *al-ḥīwār al-mutamaddin* [The civilized dialogues]. 29 November 2006. Accessed 4 December 2009. www.ahewar.org/debat/show.art.asp?aid=67136

Abdul Jawad, Jamal. *al-Ḥarakātul-'Islāmīyati jādatun fī tabannī-d-dīmoqrāṭīya wa lākinaha tafshalu wa tatahāwanu fī taṭbīqihā ḥāl wusūlihā li-l-sulṭa* [The Islamic Movements are serious about building democracy; however, they fail and compromise in its application in regard to acquisition of power]. *ash-Sharq al-'Awsaṭ* [The Middle East]. 17 November 2003. Accessed 29 November 2009. www.aawsat.com/details.asp?section=17&article=203492&issueno=9120

Abu Haniya, Hasan. *al-Ḥarakātul-'Islāmīya wa at-taḥāwul ad-dīmoqrāṭīya* ([The Islamic Movements and the democratic transformation]. *al-Ghad* [Tomorrow]. 30 November 2006. Accessed 20 November 2009. www.alghad.com/?news=93444

Al-Jazeera. *al-Ma'rifa, al-'aḥzābus-sīyāsīyatu fīlwaṭan al-'Arabī* [The political parties in the Arab homeland]. Prepared by the Department of Research and Studies. *al-Ma'rifa* [Knowledge]. 10 March 2004a. Accessed 15 November 2009. www.aljazeera.net/NR/exeres/E166720D-22B9-4708-B1AE-18C42A16C1FC.htm

Al-Jazeera. *al-Ḥarakātul-'Islāmīyatu wa-d-dīmoqrāṭīya: al-muwāqifu wa-l-mukhāwiful-mutabādilatu* [The Islamic Movements and democracy: mutual attitudes and concerns] *al-Ma'rifa* [Knowledge]. 10 March 2004b. Accessed

15 November 2009. www.aljazeera.net/NR/exeres/D6812447-D83A-4B7E-AB84-7A6FAF2B8C1B.htm

Al-Mu'min, Iqbal. *ayhuma 'akthar ṭā'ifīya Ādam 'am 'Iblīs?* [Which of the two is more sectarian, Adam or Satan?]. *al-kātib al-'irāqī* [The Iraqi writer]. 19 August 2009. Accessed 5 December 2009. http://iraqiwriter.com/iraqi-writer/iraq/8765441.htm

Azam, Ahmad Jamil. *murabba' al-ḥarakāta-l-'Islāmīya* [The square of the Islamic movements]. *Al-marSad al-'Arabii lil-islaaH wa ad-diimakrati-iya* [The Arabic observatory of reforms and democracy]. 31 August 2006. Accessed 8 November 2009. www.awrd.net/look/article.tpl?IdLanguage=17&IdPublication=1&NrArticle=380&NrIssue=1&NrSection=17

Bakhtin, Mikhail Mikhailovich. "Discourse in the Novel." In *The Dialogic Imagination*, edited by Michael Holquist, translated by Caryl Emerson and Michael Holquist, 259–422. Austin: University of Texas Press, 1981.

Bakhtin, Mikhail Mikhailovich. *Speech Genres and Other Late Essays.* Edited by Caryl Emerson and Michael Holquist, translated by Vern McGee. Austin: University of Texas Press, 1986.

Briggs, Charles L., and Hallin, Daniel C. "Biocommunicability: The Neoliberal Subject and Its Contradictions in News Coverage of Health Issues", *Social Text* 25, no. 4 93 (2007): 43–66.

Ghalioun, Burhan. *ad-diimakratiiya al-'Arabiiya wa bu'bu' al-ḥarakaat al-Islaamiiya* [Arabic democracy and the bugaboo of Islamic movements]. *Al-Jazeera Net*, 20 December 2003. Accessed 8 November 2009. http://critique-sociale.blogspot.com/2003/12/blog-post_20.html

Irvine, Judith T., and Gal, Susan. "Language Ideology and Linguistic Differentiation". In *Regimes of Language and Ideologies, Politics and Identities*, edited by P.V. Kroskrity, 35–83. Santa Fe: SAR Press, 2000.

Oktar, Lütfiye. "The Ideological Organization of Representational Processes in the Presentation of *Us* and *Them*", *Discourse & Society* 12, no. 3 (2001): 313–46.

Siddiqui, Iqbal. "Dangers of 'Democratizing' Islam and Muslim Political Discourse", Media Monitors Network, July 21 2003. Accessed 22 May, 2018. https://www.mediamonitors.net/perspectives/dangers-of-democratizing-islam-and-muslim-political-discourse/

Sidorkin, Alexander. "Lyotard and Bakhtin: Engaged Diversity in Education", *Interchange* 33, no. 1 (2002): 85–97.

11

RECLAIMING THE SACRED

The Bengali Muslim community's quest for a *jatiya* identity

Epsita Halder

> To construct *jatiya* life, Muslim boys should be reminded of the glory of their Islamic past in the tales they heard in the family, in school textbooks, at religious institutions, at public gatherings. The *jati* that does not have its history, does not have any honour. To save our *jatiya* existence, we need to develop Bangla, our language.
>
> "Bangbhasha O Muslim Sahitya", Ahsan Ullah
> (Bangla Language and Muslim Literature)
>
> Delivered at Jessore-Khulna Siddiquiya
> Sahitya Samiti Session 1918

Jatiya became the pivotal expression in the Bengali Muslim public sphere from the late nineteenth century to define an identity-in-difference as a counter-narrative to Orientalist-Occidentalist discourses and Hindu nationalist ideas. When the contemporary Urdu-speaking Muslim intelligentsia of north India started debating the definition of Muslim collective identity, they attempted to express the notion of collective and pan-Islamic belongingness through words like *qaum*, *watan*, and *millat*. While the Urdu-speaking intelligentsia had so many terms to designate Muslim identity within pan-Islamic and national modernist discourses (Jalal 1999), Bengal Muslims had only the expression *jatiya* to mark different dimensions of belongingness and collective consciousness, no matter how multifarious and fractured that was. *Jatiya*, henceforth, had several interpretive contexts – not a unilinear one – to understand Bengali Muslims' search for a collective identity which had pan-Islamic, national, and regional fervours. Also, *jati*, then, became the most powerful signifier to denote both racial and religious identities – conflating race with religion – important factors for defining difference.

In this chapter, I attempt to unpack several contexts within which, starting in the late nineteenth century, *jati* and *jatiya* were formulated in the Bengali Muslim public sphere by several groups – antagonistic to and resonating with each other to form a set of common denominators for a secure identity to finally arrive at a moment when the trans-regional and the local converged. The issue over *jatiya* emerged in discussions of literary platforms and social associations, making language-literature a political question for Bengali Muslims. The reclamation of a Sanskritized standardized Bangla as the marker of Bengali Muslims' *jatiya* identity became an auxiliary element in the search for pan-Islamic political identity, a concerted effort to separate Muslim *jatiya* literature from the religious and narrative texts hugely proliferating in Muslim Bengali in popular cheap print formats. This article also entails a study of the dialogue between literary activities and political events by looking at literary expressions arising out of religious essence and being communicated through regional idioms to carry the thrust of political orientation and solidarity.

I attempt to engage with such deliberations of the Bengali Muslim literati by reading two genres – biography and history – that they produced in the late nineteenth and early twentieth centuries to mark and mould collective identity in religious and political terms. In this effort, I look beyond binary parameters of religious/political, modern/traditional, and elite/popular to engage with the Bengali Muslim community's struggle for linguistic identity, modernity, and political solidarity.

The search for *jatiya*: history-biography-literature

They poisoned Hasan and killed Kasem thirsty . . .
A million soldiers slayed a solitary Husayn
We will remember bloodstains
The son of your own disciple killed your progeny.
Our hearts shattered, broke into many.
On the fields of Hasar, revenge could be sought.
But we will forgive the enemy; blessings for the *ummah* will be brought.

(*Asheq-e-Rasul*, Daal Ali 1908)

This section from *Asheq-e-Rasul*, a long praise poem on the Prophet by Mohammad Daad Ali (1852–1936),[1] a practising Sufi of Bengal, brought alive the life of Muhammad as history, biography, and poetry at the same time. This generic expansiveness was a curious attribute of the Bengali Muslim *jatiya*, as the authors were transcending generic boundaries to establish the Prophetic theme as the *jatiya* theme. In this process of forming *jatiya* – history, biography, and literature – based on prophetic connections, the Bengali Muslim intelligentsia negated emotions connected to lived religious traditions and reformulated genres to separate the scriptural from the sensory, in a sense saving history from fictional fabrication.

Generic polyvalence marks *sirat* – sacred biography – as having the potential of scripture, narrative, and history all at the same time. *Sirat* during the medieval and the early modern periods offered emotional cues to the audience to identify with Islamic narrative imaginations and through that to *tauhid* (oneness of Allah). *Sirat* authors like Sayyid Sultan (*Nabi-vamsa*, 1560), Sheikh Chand (*Rasul-Bijoy*, 1700s) and Heyat Mamud (*Ambiya Bani*, 1758), along with their transactions with local high Vaishnavite, Natha and Buddhist ideals were also open to a more fluid *jungnama* (narratives based on the military exploits of Islamic heroes historical or mythological), while retaining Islamic theological concepts in their *sirat*.

The Prophet, with the advent of Islamic reform in the mid-nineteenth century, was projected as the icon whose life was to be emulated by the newly emerged and emerging readership. To accomplish that, a whole gamut of biographies was written and printed on the lives of the Prophet, his *ashabs* (four companions of the Prophet, who became the *khalifas* after his death – Abu Bakr, Umar, Usman, and Ali) and the sacred characters of *ahl-ul-bayt* (the family of the Prophet – daughter Fatema, son-in-law Ali, and grandsons Hasan and Husayn).[2] To understand the multitude of the connections between pan-Islamism, identity as Bengali Muslims, reformist values of Islam, and choice of a regional language to articulate identity and modernity, I will engage with the claims of biography and history charted out by the literati connected with two periodicals – *Sudhakar* (started in 1889) and *Mohammadi* (started in 1903).[3] As *Sudhakar* and *Mohammadi* are concerned, we will look at the aesthetic-poetic questions arising in Bengali literary culture as integral to the processes of identity formation. Through the engagement of the Bengali Muslim literary and religious intelligentsia writing standardized genres in Sanskritized Bangla about Islamic history and sacred characters, the Muslim community strove to finally attain political identity and solidarity by forming a *jatiya* identity. This was articulated by Shaikh Abdur Rahim, who said: "Our *jatiya* history is confined in the fort of Arabic language. . . . We are completely ignorant about our *jatiya itihas*. No *jati* can progress without knowing its past, this is an unquestionable truth" (Rahim 1910).

The nineteenth-century turn towards Prophet-centric piety, with the Prophet as the ideal man, marked the core values of *jatiya*. 'This-worldly' Islam (Robinson 2004: 47–58)[4] attempted to nullify the effects of the charismatic *pirs*, who were the carriers of a more physicality-based

Islam and the traditional *mullahs*, who simplified Islam with little or no exposure to original scriptures. Since the beginning of the nineteenth century, reformist Islam had struggled to bring the community back to the fundamentals of Islam by marking and purging out 'un-Islamic' elements, such as *shirk* (polytheism) and *bid'ad* (innovation), that seeped in through close proximity with local religious-ritualistic imaginations and practices. To accomplish Muslim religious reform, the era of the Prophet and after his death, that of *kholafaye rashidin* – Abu Bakr, Umar, Usman, and Ali – were idealized, which manifested in claims of *jatiya* by reclaiming the Prophet and the Islamic caliphate. The Prophet's life not only became *the* life, but now was authenticated by/as *hadis* as the ideal template to follow for Muslim masses. The interconnected political and personal issues of his life – the Islamic state (caliphate) and inheritance – became the most crucial factors to resolve the crisis of *ummah* (the community of the Prophet) as the modern political system for the making of the collective – *jatiya*.

But such attempts to write history by the Muslim intelligentsia were not intended as a counter-narrative to Orientalist or Occidentalist discourses on Islam, although colonial rule did trigger and aggravate certain generic and ideological formulations. Rather, the attempt to write history can be traced to the conceptualizations of the Prophet in reformist politics and theology in eighteenth-century north India. Shah Wali Allah (1702–60), a *muhaddith* (*hadis* scholar) and reformer of the time, minimized the divine status of Muhammad to interpret the superiority of heart and mind as a historical medium (Harlan 2008). Despite the acute antagonism over theological issues between Hanafi scholar and jurist Shaikh Ahmad Shirhindi (1564–1624) and Shah Wali Allah, the two were on the same page in regard to the historical model of the Prophet's life and the absolute authority of divine law (Robinson 2004: 19–20, 31). The ideal of Muhammad was again reaffirmed by Sayyid Ahmad Shahid (d. 1831), the founder of reformist *Tariqah-i Muhammadiyah* in the later period, who propagated a scriptural model of Muhammad with his humane and historical attributes (Robinson 2004: 37). Though the interpretation of reformist Islam gradually took different shapes and forms, the figure of a scriptural and historical Muhammad as the ideal ethico-political leader of the community remained constant in the writings of the Muslim intelligentsia from the mid-nineteenth century.

Reformist movements – initiated and spread in the remotest parts of Bengal by travelling *khalifas* (chosen descendants from disciples to maintain the legacy of the master) of *Tariqa-i-Muhammadiyah*, local Mir Nisar Ali or Titu Mir (1782–1831), a disciple of Sayyid Ahmad Shahid and the Farazi leader Haji Shariatullah (1781–1840) – brought a sense of belonging to the Islamic *ummah* and the proper way of doing it by separating it from the local ritualistic customs.[5] Such endeavours energized print culture by proliferating religious treatises on *iman* (faith) and *ada'ab* (the right way of practising Islam) and also by bringing a scriptural turn to the narrative tradition. Religious reform opened up a new network of reformed *madrasa* education and started producing a new generation of *ulama* with proper scriptural knowledge which differed qualitatively from the *sabiqi mullahs* (the pre-reformist old generation of *mullahs*) against whose version of Islam the whole reformist movement was targeted. Though late to respond to the forces of modernity in comparison with its Hindu counterpart, the Muslim community was now becoming a part of modern colonial institutions to form its own collective identity. The neo-traditionalism of Afghani, which was all about the tugs between the between orthodox traditionalism and pure-western, sought modern values within Islamic tradition (Hasan 13; Keddie 3). This, when received by the learned Bengali *munshis* and *maulavis* after Afghani visited Calcutta in the early 1880s, created an urge to realize and articulate collective identity, taking cues from the Islamic past (Ahmed 83, 110), *Sudhakar* being the initiator of such efforts. With Sayed Ameer Ali's *Spirit of Islam* and *Life of Muhammad* well received by neo-traditionalists, *Sudhakar* emphasized an irreconcilable gap between Christians, Muslims, and Hindus as the reaffirmation of pan-Islamism and as a counter-narrative to Orientalism, Christian missionary activities, and Hindu nationalism. But that did not discourage them from imbibing a Bengali cultural register that was saturated with essential Hindu nationalistic imaginations and affect.

If we look at the broader historical outline, the new sense of belonging to the *ummah* was realized by the Muslim masses of the 'tropical swamps' of Bengal during the anti-British *jihad* movement on the north-western frontier during the first wave of the reformist Tariqa movement. While the Wahhabi trial (1864–71) succeeded in reducing the insurgent energy of the common masses, the first-generation Bengal Muslim intelligentsia emerged with its capacity to connect to trans-territorial events. For instance, the Greco-Turkish War (1897) and the construction of the Hijaz railways by the Ottoman sultan to connect Medina (1908), in various ways, strengthened their realization and articulation of pan-Islamism.

Sudhakar was first published as a weekly journal in 1889, through the collective efforts of Muhammad Reyazuddin Ahmed, Pandit Reyazuddin Ahmed Mashadi, Maulavi Meyarazuddin Ahmed, Shaikh Abdul Karim, Mozammel Hoque, and Dr Habibur Rahman. It was heralded as the "first *jatiya* journal of the Muslims" by Sayyid Emdad Ali (Qadir 1966: 47). *Sudhakar* signalled a paradigmatic shift by creating a literary network with an effort to standardize literary genre and language, invite new forms of patronage, and create a new readership as the marker of *jatiya* identity that was both Muslim and Bengali. *Jatiya* was formulated as essentially derived from scriptural sources and inherently regional. As a counter-narrative to the Urdu-speaking elite discourses of Bengali authors such as Syed Ameer Ali and Nawab Abdul Latif[6] about modern Muslim institutions and the public sphere, the Bengal Muslim intelligentsia had to build their own institutions and literary systems to identify with and consolidate a *jatiya* for the Bengali-speaking Muslim community. This endeavour marked Bangla as the ideal language to articulate *jatiya*, which was, at the same time, a reclamation of Bengaliness denied to the Muslim community within the hegemonic Hindu nationalist history.[7]

As an issue of *Sudhakar* said: "Newspapers, journals and periodicals are the condition for *jatiya* progress . . . Bengali is the language of Bengal, we are the Muslims of Bengal. We need an ideal newspaper where politics, sociology, economy, literature, science, history, nature, everything will be included. The glory, grace, prowess and knowledge of the Muslims of the past would be shown brightly to everybody. We will place our demands and grievances to the British government through this paper. Whatever hurts *jatiya* sentiment and interest will be taken care of here" (*Sudhakar*, 23 Kartik 1296 [1889]). With this agenda of creating *jatiya*, *Sudhakar* stood for a network of literary activities separating itself from the popular cheap print culture (now seen as unfit in terms of both genre and language to hold Muslim *jatiya* essence) by forming modern genres in the standardized language paradigm of Bengali. Not only was the interface between Islam and Bengali language, in the *Sudhakar* network, facilitated by authors with the ability to channel Islamic themes into modern genres, but also the participation of Islamic missionary preachers like Munshi Mehrullah (1861–1907)[8] and Shaikh Zamiruddin (1870–1930) energized the processes of modernization and Muslim identity formation. Also, the alliance between *pir* Abu Bakr Siddique of Furfura and the religious intelligentsia of *Sudhakar* in the 1920s was instrumental in enlarging the literary network which revitalized print, popular piety, and the crusade of *anjumans* reaching out to the masses with a message of social and religious reform. Such print activities in Bengali-speaking, aligning with religious authorities, replaced the Urdu-speaking preachers of the first wave of reformist activities (W. Ahmed 1883: 384) coming from north India. These affirmed the case for the Bengal local, involving and initiating the Bengali-speaking masses into the values of Islam through literature.

Mohammadi, the longest-running periodical of the time in Bengal (1903–8: monthly, 1908–22: weekly, 1922–7: daily, 1927–68: monthly), was a witness to sixty-five years of iconic events that shaped the contours of history and politics of South Asia. The literary and discursive contributions of its editor Maulana Muhammad Akram Khan became symbolic to the ideational and political transformations affecting Bengal, such as the proposal (1905) and annulment of the partition of Bengal (1911), the foundation of the Muslim League (1906), the Khilafat movement (1919–22), the emergence of the Bengal Provincial Muslim League (1912), and the merging of a regional Krishak Praja Party and the Muslim League (1937) addressing the demands of Muslims and East Pakistan, which changed the relationship between north-Indian political activists/intelligentsia and their Bengal counterparts. Muhammad Akram Khan was associated with all these political and cultural activities by shifting and reformulating his ideological positions to finally conclude that the creation of Pakistan could do much[9] to strengthen the political orientation of the *anjumans.* Politics and religious sensibility coalesced as the motto of *Mohammadi* discourse when Akram Khan's compatriot Muhammad Maniruzzaman Islamabadi (1875–1950) stated that the desire for freedom was an essential part of *iman* and according to Islam it was a sin to not to strive for it (Islamabadi 1927: 3). What emerged in the literary activities of *Sudhakar* – a sensibility of essential racial/*jatiya* separation from the Hindus – reached its full realization with Maulana Akram Khan's utterance "When it comes to *jati* or nation, Muslims are committed to *swatantrata* ['difference/distinct']" (Anisuzzaman 1969: 214), and with Akram Khan as the secretary of Anjuman-i-Ulama-i-Bangala (The platform of the religious intelligentsia of Bengal), the *anjuman* culture of Bengal entered the domain of politics by taking a clear religious reformist and anti-British stance. With Akram Khan, *Mohammadi* took a stance against British colonial power and against the autonomous secular literary paradigm initiated by *Muslim Sahitya Samaj* and its corollary periodical *Sikha*, charting out a political-literary paradigm for Muslim *jatiya* starting in the 1920s. This marked support for Pakistan's formation and positioned the Prophet as the socio-moral ideal in *jatiya* literature. This was also a moment when, rather than a rigid opposition to Urdu, the presence of the Perso-Arabic-Urdu registers in Bengali was thoroughly discussed, theorized, and affirmed while talking about Bengali language, as it were.

To understand the emergence of the national between pan- and regional Islamism since the Khilafat movement that awoke the Bengali Muslim community in the name of the nation, we need to see how *jatiya* took different shape and form. *Jatiya* started being sustained by the more politically inclined religious intelligentsia, such as Maniruzzaman Islamabadi and Maulana Muhammad Akram Khan, in terms of religion and language.

Ideals about the pure literary and the didactic differed individually even within collectives like *Sudhakar* and *Mohammadi* and changed over time. Muhammad Reyazuddin Ahmed (1862–1933), a regular contributor to *Sudhakar*, established *Islam Pracharak* (1891) with an agenda of reviving the past glory of Islam and prioritizing the ideals of *Qur'an-hadis*, like all Sudhakarists. Reyazuddin Ahmed was extremely critical of *Mihir*, established by the prolific editor of *Sudhakar*, Shaikh Abdul Rahim (1859–1931), which was "not about religion" but was "about literature and history" (*Mihir*, April–May–June–July, 1893). According to critics, the weekly *Sudhakar* and the monthly *Mihir* merged into *Mihir O Sudhakar*, which "struggled its best to organize the Bengal Muslim community as a collective and revitalize them with the ideal of wisdom and renunciation, knowledge and literature" (Qadir 1966: 52). While these two distinct lines of thought came to contradict each other on the definition and generic formulations of *jatiya* literature, Reyazuddin Ahmed and Shaikh Abdur Rahim collaborated in translating Afghani's book on Islamic theosophy and Maulana Abdul Haq's *tafsir* in two volumes of *Eslam Tattwa* (Published in 1988, Calcutta, Aswin 1295BS) in Sanskritized Bengali.

What was shared between the overtly didactic and the literary forms of *jatiya* was an impulse to standardize biography-history in Sanskritized Bengali by eliminating the imaginative and the non-historical embedded in existing popular literature. Muhammad Reyazuddin Ahmed, Shaikh Abdur Rahim, and Shaikh Abdul Jabbar (1880–1917) vehemently critiqued the supernatural elements that had intruded the popular biographies of Muhammad through Islam's close proximity with local aesthetic and religious systems. They worked hard to give Muhammad's life a shape of historical reason, basically by taking biography close to history, by making biography history, while curbing the fluidity of the narrative and mythological imagination. Within this, both Shaikh Abdur Rahim and Shaikh Abdul Jabbar could not disregard the aesthetic-poetic allure of Bengali Sanskritized prose, so integral to the literary imagination of the time, but they sanitized it with the presence of the Prophet and the prophetic.

> To date, *sirat* has been written from the imagined episodes of *maulud sharif* or translating from pseudo-*hadis*, or based on misconceived biographies by Hindu or European writers. In spite of getting religious or moral teaching, one gets into the emotion of polytheism and starts following un-Islamic elements as Islam.
>
> Shaikh Abdul Rahim, *Hajarat Muhammader Jiban O Dharmma Niti*, 1887, Muhammadpur, Basirhat

All narratives connected to performative rituals, such as the celebration of the birth of Muhammad in *maulud sharif* and commemoration of the deaths of his grandsons in *muharram*, were given rational shape, minimizing the intercessory affect towards the sacred characters and marking the impact of *shirk* as a condition of that affect. *Sudhakar* was the pioneer in this task in standardized Bengali.[10] *Milad*, *urs* (commemoration of the day of death of the *pir*), and *muharram* were critiqued by the reformist religious authorities writing about the pure form of Islam in new instructional and behavioural manuals, which were adapted by members of the intelligentsia who wanted to restore religious essence as the core of *jatiya*. Reformed *maulud sharif* was being written by the reformed *ulama* and the literati, Hasan and Husayn were reconnected with the Prophetic life, and the battle of Karbala was reclaimed as a sacred reason for sacrifice. The overall intense effort marked a turn to the scriptures by weeding out the 'false' and placing the Prophetic life in connection to the *ashab* and the family. So, once the Sunni caliphate as the ideal form of religiosity and polity was affirmed in the community's political imagination, the Shia intercessory claim over *pak panjatan* was negated in the process. Thus, along with the creation of the Bengali Muslim community as a whole, began the fixity of sectarian identity as the outcome of the Sunni reformist effort. The Shias became the 'other' in the Muslim community.

Writing *jatiya* history-biography: the truth of the sacred

> Do these young students of *madrasa* know what the *history* of Islam is? What exactly is the *historical biography* of the Prophet? Do they know Islamic theology? Are they well versed in *tafsir* and *hadis*?
>
> *Islam Chitra O Samaj-Chitra*, Maulavi Shaikh Abdul Jabbar, Mymensing 1913 (emphasis mine)

In the Islamic worldview, methodologies of writing objective history connected to jurisprudence and polity had strongly been debated through the ages, with the indispensability of the life of Muhammad remaining the primary and constant concern. Ahmad ibn Hanbal (d. 241/855) said:

> As regards traditions from the Messenger of God on what is allowed and what is forbidden, on what is laid down and what is decreed, we are tough on the *isnad* (the individual narrator), but for traditions of the Prophet on the virtues of certain actions and what does not prescribe or proscribe, we go easy on the *isnad*.
>
> (quoted in Bray 2006: 20)

To demarcate between the status of the *hadis* and other forms of historical narration – say, of *khabar*, *muhaddiths* emphasized the flawlessness of narration, which means an impartial untreated form of pure transmission.

Khabar, the other kind of narration, was predicated upon imparting a structure, a trace of mediation, valid as a form of narration depicting the truth. These two forms, *hadis* and *khabar*, as Julia Bray showed in her discussion on the culture of representation in medieval Islam, were contrasted with fictional narration like *Alif Laila* and other poetic works. Though difference between the truth and imaginative invention was thoroughly marked, fiction and history shared techniques and strategies for the composition of their discourses.

The Islamicate world, with the proliferation of Sufism and the more 'massy' Sufi narratives, did not encounter questions of historical authenticity that upset classical and medieval Arabic traditions. The Sufi narrative was not a search for history; rather, it was about the pleasure of the narrative as the secure substitute of the Qur'an and the *hadis* that were unavailable to the masses. In the vernacular mind, rather than *isnad* – the flawless reporter – emplotment, characterization, and narrative sequence were more important to the tales of sacred historical or quasi-historical heroes. In the late nineteenth century, in the available multisource of history encompassing the *hadis* and history books by the Oriental-colonial administrators and modernist English-Urdu speaking Islamic intelligentsia, it is worth noting how different social groups in Bengal were receiving and responding to this polyvalence. The *Sudhakar* group showed enormous ability to receive the already available scholarship on the life of the Prophet and Islam and a form derived from contemporary middle-class narrative imagination of Hindu cultural nationalism. From Shaikh Abdur Rahim's citations in *Hazrat Muhammader Jibon O Dharmmaniti* (1887), we get an all-inclusive arc of Perso-Arabic scriptural and historical sources (*khabar*), European books, and also Sayyid Ahmad Khan and Syed Ameer Ali, with which Rahim defended both the Prophet and Islam.

Mecca and Medina as devotional centres had been the truth in the Arabic or Persian narratives and referred to in medieval and early modern Bengali texts. But what new was their position in the Bengali devotional landscape as real spaces, their spiritual-scriptural status offering a direct connection between local and pan-Islamic values. The history of these two sacred places also offered readers a sacred experience, equivalent to the experience of Haj (pilgrimage). The logic of life that emerged through these biographies and history narrated by the *Sudhakar* group shifted the language of traditional piety to contemporary rationalism, redefining the sense of spiritual surrender to the sacred. The supernatural and the ritualistic were denied and opposed. Different generic strategies were deployed, placing the narratives within the framework of the rational-historical. Shaikh Abdul Jabbar, in his *Madinasharifer Itihas* (1314/1907), like many other authors of history, detailed the family tree of Muhammad both from the paternal and the maternal sides to place the Prophet historically. Shaikh Abdur Rahim, in *Hazrat Muhammader Jiban O Dharmaniti* (1887), cited references from *hadis* and British sources and used photographs, diagrams, and cartographical images to bring his subject closer to the modern discourses of the time.

Yet, there is the recurrent impossibility of completely abnegating the form of popular piety so embedded in the broader reception of *hadis* and different mythological variants of the scriptural texts. At times, authors of history ended up historicizing the supernatural itself while distancing themselves from a body of texts in popular print whose basic tenets of piety were predicated upon the supernatural or the miraculous. To fabricate the past, Abdur Rahim began with the Arab country of his time and discussed recent situations in Mecca and Medina in present tense and real time. While an archival photograph of the Hera cave was inserted, Abdur Rahim added scientific justifications to the oracle that brought about the *shab e-miraj* (revelation of Allah's knowledge to Muhammad), which Gibbon in his book attacked as supernatural. The attempt to bring the extra-rational into the framework of history can be seen in various other biographies and history books, which only affirms mutual seepage between an elite *jatiya*-modernist stance and a more popular form of piety, both instrumental in creating *jatiya* acceptable to different layers of the community. In the advertorial in the beginning of Shaikh Abdul Jabbar's *Medinasharifer Itihas*, Maulana Abdul Haq referred to an authentic history book by a Medinite, Sayyid Nuruddin Ali Begh Sayyid Afifaddin bin Ahmad Husaini Samhudi (named *Jwazul Qulub* in seventeenth-century Delhi), which became a major source not only for Abdul Jabbar but also for other writers of biography, history, and literature dealing with sacred lives. While attempting to chart out the first book of history of Mecca and Medina in Bangla, the author could neither eliminate nor minimize the devotional excess that the land of Medina stood for in the collective psyche. Abdul Jabbar's biography of Muhammad had the picture of 'Kaba Masjid'a with its four sides signifying four *mazhabs* (Hanafi, Shaifi, Makeiki, and Hambali – interpretive law schools of Sunni Islam which gave rise to further sub-sectarian divide with the names of the Sunni lawmakers). In the introduction, poet Kaykobad emphasized the historicity of the life of the Prophet with the phrase *jiban-brittanto* ('chronicles of life') standing out in bigger font, the visuals indicating the *mazhab* used disregarding chronology.

Thus, the historical and the rational were fluid enough to stretch and connect several layers of Muslim psyche, operating in different registers of reason and historicity. What is significant here is the reciprocal and almost supplementary connection between an evidentially proven history and the narrative expansion of any historical theme. This was mostly celebrated by Abdul Jabbar, who chose Kaykobad the poet, rather than some religious rationalist, to write an introduction to his biography. While continuously referring to sections from the *hadis* repertoire to authenticate every event and utterance, the narrator remained anecdotal, exclamatory (with an array of question and exclamation marks), and a bit inclined to the lyricism of literature already in circulation among Hindu litterateurs.

Reception, refutation, and reclamation: *jatiya* literature

When Sir Sayyid Ahmad Khan, a proponent of Western education, wrote *A Series of Essays on the Life of Mohammed from the Original Sources*[11] as a counter-narrative to William Muir's Orientalist-colonialist *Life of Mohamet* (1861), two purposes were served. Sir Khan made a paradigmatic break in the way the life of the Prophet was being written so far in the traditional *sirat*. The text was massively influential in formulating history and biography not only on the elite Urdu sphere, but also on vernacular cultures like Bengali.[12]

From Sir Sayyid Ahmad Khan to the major and minor Muslim writers of Bengal, everyone attempted to solve the same riddle – 'What are those facts in reality?' – and along the way produced the genre of history. Before Shibli Nomani (1857–1914) could finish his *Sirat-un-Nabi*,[13] Syed Ameer Ali (1849–1928) published his biography of the Prophet entitled *The Spirit of Islam, Or the Life and Teaching of Muhammad* (1874). This book was a direct counter-narrative to Muir's. Written in English, it addressed Muir's readership. Syed Ameer Ali employed two genres, biography and history, along with his other writings on *fiqh* (Islamic law). In *The Spirit of Islam*, he put the biography of Muhammad and the development of Islamic theosophical, social, and political ideas within the framework of a single book to illustrate biography's ability to function as both scripture and history.

Girishchandra Sen (1835–1910)[14] stood between Syed Ameer Ali and the *Sudhakar* network to attempt the life-stories of Muhammad and his grandsons in Sanskritized Bangla. But his agenda was different from Ameer Ali's. Girishchandra Sen was the first non-Muslim Bengali to engage with a plethora of original Arabic and Persian scriptures and create a discursive corpus of Islam in Bangla.[15] His exploration of *Qur'an-hadis* and his coinage of *mahapurush* ('a great man') to designate Hazrat Muhammad inspired another generation of authors like Shaikh Abdur Rahim and Shaikh Abdul Jabbar of *Sudhakar*. Girishchandra's attempt in *Mahapurush Charit*, in which he described all the prophets of Islam in three volumes and the biography of the Prophet in three volumes (1886–7),[16] was incomplete without a separate book on the life-stories of Imam Hasan and Imam Husayn (1901).[17] It is notable that Girishchandra was the first author to have attempted a separate 'biography' of Imam Hasan and Husayn and the *ashab* in Bangla.[18] Though an active member of Brahmo Samaj, Girishchandra stood outside of the colonial-Christian missionary-Brahmo missionary-Hindu hegemonic construction of the Muslim community and Islam and created a connection between Islamic scriptures, devotionalism, and the Bangla-speaking readership.

Sudhakar's Reyazuddin Ahmed (1862–1933) wrote *Hajrat Muhammad Mostafa* (1924), *Hajarat Fatema Zohra (Ra) er Jiban Charit* (1928), and *Pak Panajtan* (1929); Shaikh Abdur Rahim wrote *Hajarta Muhammader Jibon O Dharmmaniti* (1887), *Islam Itibritta* (1910–1); and Shaikh Abdul Jabbar wrote *Makkasharifer Itihas* (1904), *Islam Chitra O Samaj Chitra* (1907), *Medinasharifer Itihas* (1907), and *Hajarater Jibani* (1914).

In all the biographies and books of history, the era of the Prophet and the caliphate was presented as a golden age. The dates of publication of the separate biographies of the *ashab* coming from various districts corresponded to the increasing impact of the Khilafat movement in *mufassil* Bengal, which only confirms the increasing reliance on the caliphate as the bearer of Islamdom by deriving sanction from the Prophet and the ability of the *ashab*.

> The more there is discussion of the life-story of this great **man**, the more the effect of sin will be reduced, and men who deviated from their path will move towards moral goodness.
>
> Introduction by poet Kaykobad to *Hajrater Jibani*,
> Shaikh Abdul Jabbar, 1913 (emphasis mine)

The very emphasis on the greatness of a 'man' named 'Hajarat' stood as a self-sufficient self-explanatory signifier in this new turn in biography/history by the Bengali Muslim religious intelligentsia.

Shaikh Abdul Jabbar's Prophet became a man, shedding his Prophetic *darood* – *a.s./pbuh* in the text. There followed massive public debate to critique or justify this absence of *pbuh*, giving rise to various ideologies in which religiosity was modernized, presenting devotion as an inward quality, a capacity of mind – in this-worldly Islam – which did not need explicit religious utterance. With *darood* gone, these authors of biography and history envisaged a deliberate tryst with modernity, by bringing the sacred narrative close to the domain of modern literary genres and making sacred lives (especially Hasan and Husayn) perform in real time and real space as historical characters.

It is a futile attempt to generically demarcate history, biography, and literature in the understanding and formulation of *jatiya*, which was the essence of Islam qualifying all the forms as *jatiya* literature. For both *Sudhakar* and *Mohammadi*, this essence was called *dharmabhab*, any deviation from which provoked disqualification from *jatiya*. Mir Musharraf Husain was attacked heavily by the *Sudhakar* group for not being able to offer *jatiya* in his rendition of the battle of Karbala in the Sanskritized prose narrative *Bishad Sindhu* (Ocean of Sorrow, the first Bengali prose narrative on this theme, published from 1887 to 1891), as the battle of Karbala was seen as key to creating *jatiya bhab*. Without *jatiya adarsha*, no literary rendition of the sacred was held legitimate as *jatiya* literature by *Sudhakar* and *Mohammadi*. Two narratives on the battle of Karbala, *Bishad Sindhu* and *Mahasmashan* (The Great Cremation Ground, a long lyrical narrative published in 1904) by Kaykobad, attracted much antagonism from *Sudhakar* and religious authorities for retaining "un-Islamic" emotions and elements (Saiyyad Emdad Ali 1917; Muhammad Reyazuddin Ahmed 1930). Only those narrative strategies that were framed to tell the stories of Islamic ideals and virtues were permissible.

Within *Sudhakar*, a curious ambivalence was inevitable too. While *Sudhakar* allied with Munshi Naimuddin in his vehement critique of Musharraf Husain, which ended up in a prolonged lawsuit, Shaikh Abdur Rahim recognized the radicalism of Naimuddin's translation of the Qur'an and other religious texts into Bangla as equivalent to Musharraf Husain's rendition of the historical battle of Karbala into literature.[19] But Shaikh Abdur Rahim was an exception among *Sudhakarists* who were less amenable to narrative indulgence in sacred texts.

Sudhakar imagined a *jatiya* in which history and biography could be termed *jatiya* literature without much generic distinction. Literature could become *jatiya* only when it was an exclusive depiction of the sacred following the scriptural sources. Starting in the 1920s, attempts were made to address the emerging autonomous domain of literature and to separate this literary culture from the scriptural and the religious. Bangiya Muslim Sahitya Samiti was founded in 1911 as a response to the Hindu hegemonic Bangitya Sahitya Parishat to let Hindus know about Muslims. A place of multiple views and ideas, the Samiti brought together academics, emerging socialists, and *ulama* with institutionalized education, journalists, and poets. Muhammad Akram Khan structured the activities and policies of the Samiti along the lines of religious ideals. Later on, through *Mohammadi*, and his connection to Anjuman-e-Olama-e-Bangla and various social reformist activities, Akram Khan was also able to consolidate this line of religion. The ideological conflict between self-claimed secularists of *Sikha-Saogat* and the 'Islamic' *Mohammadi* journal can be read as an iconic split between literature and religion as two ideologies in defining Muslim *jatiya*, modernity, and literariness, both claiming equal viability in defining the Bengali Muslim identity.[20]

The *hadis-tarikh* and the *Qur'an-hadis*: the two reasons for *jatiya*

Though Sudhakar struggled to eliminate the thrust of popular piety in the formulation of *jatiya*, it was not possible till Maulana Muhammad Akram Khan wrote *Mostafa-Charit*. What remained constant was religious ethics – *dharmabhab* – as the core of *jatiya* and Muslim identity. But with Maulana Akram Khan, the Bengali Muslim intelligentsia could finally insert the rationale of the scriptures in the biography of the Prophet. Though both wrote religious treatises based on the readings of Qur'an and *hadis*,[21] Abdur Rahim and Abdul Jabbar remained licentious while writing history and biographies, in the sense that they preferred narrative quality to induce emotion.

But what Akram Khan proposed and achieved was a complete reliance on the Qur'an and also on the *hadis* repertoire, screening them to choose authentic ones. He disregarded the *tarikh* and distinguished between the authentic (*sahi*) and pseudo-*hadis*. This take on religion can be interpreted as the moment of arrival to the Arabic cosmopolis, so apparent in the *anjuman* movement for Muslim political solidarity. It led to the design of an Arabic University and a Bangla language sprinkled with Arabic-Persian-Urdu vocabulary that Akram Khan institutionalized as the president of the Bangla Bhasha Unnayan Board (Board for the development of Bengali language).

This discussion remains incomplete without the inclusion of Maulana Maniruzzaman Islamabadi (1875–1950), the compatriot of Maulana Akram Khan, who wrote extensively on Islam and on Muslim excellence in India and also attempted biographies. The contributions of these men were very important to what started as the expression of *jatiya*, formed *jatiya* literature and took the shape of the *anjuman* movement in the hands of Maulana Akram Khan and Maniruzzaman Islamabadi.

The *Sudhakarists*, under the pan-Islamic thrust, wrote the history of Mecca-Medina and then Turkey, connecting their region with supranational consciousness. With gradual political mobilization during the Khilafat movement, the Bengali Muslim community was able to engage with national political platforms and another conception of history along with it. The history of Muslims in India was thoroughly attempted by Maniruzzaman Islamabadi, and, along with Persian poetry, Urdu poets Hali and Iqbal were translated on a large scale. Gholam Mostafa (1879–1964), the prolific poet and a contributor to *Mohammadi*, translated Iqbal along with writing original poetry and narratives. Within the hegemonic discourses of nationalistic political organizations, such as the Indian National Congress, during the Khilafat movement, and also the authoritarianism of the upper Indian elite Muslims in the functioning of the Muslim League, Muslim leaders and the literati of Bengal were prompted to reclaim their Bengaliness as their core identity but now with a new national consciousness. With pan-Islamic, national Indian, and regional Bengali identifications, it is a 'trichotomy' inevitable for the Muslim communities in Bengal.

Shibli Nomani, who tried to find scriptural and historical sources of Islam in India,[22] was translated by Islamabadi. In an article titled "*Non-co-operation ba Asahajogita*," published in 1920 in the illustrated monthly journal with a suggestive title *Moslem-Bharat* (Muslim India), the narrow confinement of the local Masjid was extolled to extend to the codes and conducts of pan-Islamic imagination; but the author warned his readers not to be so exclusive as to offend non-Muslims, as that would disturb the national unity of India (Abdullah al-Azad 1920: 634). Nomani tried to resolve the main anxiety of Muslims in choosing between the supra-national and the national, by saying that if one supported Khilafat, he should go against it if the caliph attacked India but should go against India if it attacked Afghanistan. It was a question of going against tyranny, he asserted, in which one had to prioritize between Muslim and non-Muslim according to need (Abdullah al-Azad 1920: 666).

Maniruzzaman Islamabadi followed the Shibli Nomani line of counter-Orientalist discourse. His agenda of writing history for Bengali Muslims was focused on articulating pan-Islamic sensibility with the national framework. Through *Bharate Muslman Sabhyata* (Muslim Civilization in India, 1914) and *Bharate Islam Prachar* (The Spread of Islam in India, 1915), he supplemented with history scriptural attempts like Munshi Naimuddin's translation of *Fatua Alamgiri* (1884–92).[23] The Khilafat movement, a joint venture of Muslims and Hindus against the British, was based on an understanding of national identity. The Muslim religious intelligentsia was able to appropriate pan-Islamic symbols to envisage a pan-Indian Muslim realization (Mitra et al. 2004: 335).

In the middle of these cultural-political forces, *Mostafa-Charit*, the biography of the Prophet by Muhammad Akram Khan, showed how – beyond the Hindu nationalist framework that *Sudhakar* responded to, and also beyond the Western paradigm of reason as charted by Sir Sayyid – could the rationalism of the Qur'an be achieved. *Mostafa-Charit* is not a linear biography of the Prophet; rather, it is an ambitious, discursive project to draw from the Qur'an and the *hadis* the justification for reason in Muhammad's life. Muhammad Akram Khan marked and used all the techniques of argument and strategies of narration that the authors of the *hadis* repertoire formulated on true Prophetic events and speech.[24]

In a stark departure from Sudhakarists' indiscriminate arrangement of references from *hadis*, *tarikh*, and *sirat*, Muhammad Akram Khan (from the very outset, in the introduction) qualitatively demarcated between scriptural texts (the Qur'an and the *hadis* repertoire) and *tarikh* and *sirat* as having two different values – the former defined as religious texts and the latter (*tarikh* and *sirat*) as descriptive narratives which might not have authentic sources (Khan 2010: 7).

Muhammad Akram Khan answered this question – if there were so many authorities in scriptural knowledge in Islam, why did such deviations from the truth of Muhammad's life occur in *tarikh* and *sirat*? He said that the authorities knew that without the validation of divine Qur'an and authentic *hadis*, nothing was attested as truth. So, it was redundant for the masters to waste time and test those narrative genres (Khan 2010: 5–7).

To posit legitimate Islamic knowledge, Akram Khan categorically dismissed forms of mediation. In the introduction to *Mostafa-Charit*, he demarcated between the truth in *hadis* and what the *rabbis* (commentators of *hadis*) deduced or speculated (2010: 15). It was clear that Akaram Khan was positing a unilinear model of scripture demarcating the authenticity of the Qur'an and *hadis* to derive the authentic form of Islam, rejecting the prevalent popular practice of considering anything written in Arabic or Persian as the truth. The ideological difference came from his *Ahl-e-Hadis* lineage.[25] *Ahl-e-Hadis*, as an offshoot of the Tariqa-i Muhammadiyah movement, had the Qur'an and *hadis* to be attained through *ijtihad* – a matter of individual understanding and reasoning – rather than following the path of *taqlid* – an unquestionable faith in the path (*mazhab*) created by the Sunni jurists. That is the reason why Akram Khan was not assured by later additions to the *hadis* repertoire as offering an authentic form of Islam. He, as a *Muhammadi* (follower of *Ahl-e-Hadis*), wanted to create direct connection with the Qur'an and the *hadis* repertoire before the Sunni *mazhabi* jurists who created the four Sunni *mazhabs* of Hanafi, Maliki, Shafi'i, and Hanbali. Without exploring sectarian reformist orientations, the ideological differences and the choice of authentic texts (to write *jatiya* history and *jatiya* life) cannot be understood. Sub-sectarian and reformist orientation should be taken as an interpretive tool to look at the multiplicity within the Bengal Muslim community which made the search for *jatiya* so varied and multifarious.

To go back to the question of linguistic identity, it might not be adequate to say that Akram Khan, the visionary leader of the Bengal Provincial Muslim League, always proposed and sanctioned Bangla as the language of Bengal Muslims after 1947. Rather, we need to look at another interpretation of Bangla, different from the *Sudhakar* network. When both *Mostafa-Charit* and *Mochhlem Banger Samajik Itihas* (*The Social History of Muslim Bengal*, 1965) had Arabic originals, with Bangla translation and notes added to the text, a differently layered Bangla took shape from this juxtaposition. The East Bengal Language Committee (1949), with Akram Khan as its president, standardized the transliteration of Arabic-Persian-Urdu vocabulary that was integral to Bangla, creating a Bangla different from that of the *Sudhakarists*, who claimed the Sankritized register. This Bengali was, for Akram Khan, a religious choice: "I call my knowledge of Bengali as *ilm-e-ladunni* [knowledge sanctioned and given by Allah]. I have taken up Bengali to express the value of *jihad*. . . . Ours is a family of jihadists. . . . But Allah has blessed me with a writing tool rather than a weapon" (quoted in Munsuruddin 1960–6: 245). Here, we should not forget the new *ulama*-oriented literary and social network which Akram Khan was close to and which became a part of his implied readership and his political sensitization.

I will end by pointing out a moment of arrival for Muslim *jatiya* in the biography of the Prophet by Gholam Mostafa. Gholam Mostafa retained Akram Khan's 'Back to the Qur'an' call and the Qur'anic values of the Prophet in *Biswanabi* (Prophet of the world, 1942). But in *Biswanabi*, the pleasure of the narrative resurfaced. *Biswanabi* synthesized Qur'anic values and the pleasure of the narrative by keeping the Qur'anic verses now in Bengali translation and endowing the events and situations with the emotive values of literature. He even named one of his chapters with a line from Rabindranath Tagore's song – "shey agun chhorhiye gelo sob khane" (That fire spread everywhere) – to signify Islam's prolific expansion during the time of the Prophet. It is to be noticed that by using Tagore's puja (devotion) song "Tumi jey surer agun lagiye dile mor prane" (you have kindled the fire of music in my heart) in the title, Biswanabi attempted to secure a place for the Prophet's biography within the established Bengali literary field of the time. The title appeared to resonate with the target audience, striving to attain metaphorical and subtle layers of religiosity for Prophet-centric piety that Tagore's songs and their abstract imagery had earned in Bengali literary modernism.

It is such synthesis that made Mostafa's *Biswanabi* the most popular biography of the Prophet among Bengali Muslims to date, with multiple editions. It also has explanations of the Qur'anic references and several conceptual elements of Islam and contextual issues, such as the role of science, and comparison between Christianity and Islam. Without looking at such confluences – between the elite and the popular, between scriptural and mythological, between the Qur'anic and the narrative, between *Hanafi* and *Mohammadi*, between literary societies and *anjumans* – no modernity and *jatiya* could be attempted to understand the processes of identity formation of the Bengali Muslim community.

Notes

1 Daad Ali was from Atigram, near Kushtiya, in eastern Bengal, and wrote and published several Sufi treatises in the late nineteenth and early twentieth centuries.

2 Contemporary characters started entering the field of biography in the early twentieth century. This list includes the Sultan of Turkey, the ruler of Afghanistan, Sir Sayyid Ahmad Khan, Mohammad Iqbal, Kemal Pasha, and local Bengal philanthropists like Haji Muhammad Mohsin, proving the community's eagerness to respond to the contemporary pan- and national Islamic political issues and social changes. But in this article I deal only with the sacred lives and times while talking about *jatiya* of the Bengal Muslims.

3 This model of enquiry is not inclusive, and the choice omits the plethora of instructional and narrative texts in Muslim Bangla that, as an effect of religious reform, were so instrumental in strengthening Islamic solidarity by creating vast literary communities. Popular print culture thrived and used older generic forms, such as long narratives in rhythmic couplets, to articulate old and new themes, the religious polemic of the community, and finally political orientations.

4 "a devaluing of a faith of contemplation of God's mysteries and of belief in His will to shape human life, and a valuing instead of a faith in which Muslims were increasingly aware that it was they, and only, who could act to fashion an Islamic society on earth. This shift of emphasis has been closely associated with a new idea of great power, the caliphate of man" (Robinson 2004: 47–58).

5 Though their positions were different in designating India as *dar-ul-harb* (the land of the enemy) and *dar-ul-islam* (the land of Islam) and the *juma* prayer, their anti-British stance obscured such differences to the British.

6 As Rafiuddin Ahmed points out, without any concern for the vernacular-speaking Muslim communities of Bengal, the Urdu-speaking *ashraf* had a narrow social base and class-specific demands as community demands (Rafiuddin Ahmed 1981: 35).

7 From historians to literati, the glorious past and Muslim invasion as the core cause of the decline of Aryan excellence was a common trope in Bengal discursive and literary creations since the mid-nineteenth century. Bankimchandra Chattopadhyay, Bhubeb Mukhopadhyay, Vivekananda, Rameshchandra Dutt, Akshay Kumar Maitra, Dwijendralal Roy, and Girishchandra Ghosh did much to inculcate this form of nationalist spirit. Historian Rakhaldas Bandyopadhyay had this nationalist undertone in his making of the history of Bengal (*Bangalar Itihas*), as did Jadunath Sarkar in his *History of Bengal*, in which he could not see Muslims bringing more than 'bloody and unrelieved barbarism an tyranny' upon Bengal.

8 Hailing from a modest background in Jessore, the illustrious Meherullah mastered Arabic-Persian-Urdu and elite Bangla, counteracted Christian missionary activities, and, by writing religious tracts and preaching sessions, strengthened the religious base of the Bengal Muslim community. In 1887, he established the first *anjuman* in Jessor, named Islam Dharmottejika.

9 Other than his discursive writings with his staunch support for the creation of Pakistan, Akram Khan also wrote poetry hailing Pakistan and its ideator Muhammad Ali Jinnah. It is surprising that while throughout his literary career Akram Khan used Sanskritized Bangla with Arabic scriptural

vocabulary, his *Pakistan Nama* (1843) was written in Muslim Bangla to reach a specific audience who were not conversant in standardized Bangla.

10 The cheap print culture, with its tremendous popularity, reached the masses to bring them out from their traditional lived forms of piety to the fundamentals of Islam. *Bedaral Gafilin* (Munshi Sameeruddin 1887), *Maydanal Ulum* (Maulavi Abdul Aziz 1891), and *Kalir Fakirer Khela O Alemganer Nasihat* (Abbas Ali 1920) were the products of such popular reformist zeal. Shaikh Abdul Jabbar (not of *Sudhakar*) did not even hesitate to confirm *Muharram* as the cause of the 1897 earthquake in his behaviour manual *Fesanaye Shor- e- Qeyamat* (1897). But this article deals only with the use of Sanskritized Bangla circulated for the middle-class reading public.

11 The last revised version of this book was published in 1912. It was then reprinted in 1923 under the supervision of T.H. Weir.

12 Muir's placing of the life of Muhammad in/as historical events thus historically proved an essentialist Orientalist notion of the Prophet and Islam. But this was a paradigmatic shift from Christian missionary efforts to impart historical evidence to place Islam and its Prophet as essentially demeaning, derogatory, and false (Balijon 1948: 88).

13 This *sirat* was completed and published by Nomani's disciple Syed Sulaiman Nadvi in 1914, after Nomani's death.

14 Hailing from Dacca district, Girishchandra Sen moved to Mymensingh, where he converted to Brahmoism. Under the direction of Keshabchandra Sen, the chief of *Nababidhan Sangha* of Brahmoism, Girishchandra started learning Islamic knowledge systems in Lucknow. After learning Arabic, Persian, and Urdu, he translated the Qur'an and the *hadis* and also other scriptural manuals and *fiqh* texts – Sufi treatises, along with the biographies of the Islamic prophets, Hazrat Muhammad, Imam Hasan, and Husayn.

15 Munshi Naimuddin's translation of the Qur'an began publication in 1887, when all the three volumes in twelve parts translated by Bhai Girishchandra Sen were already published (1881–5), Ali Ahmad, pp. 379–394.

16 *Mahapurush Muhammader Jiban Charit*, Volumes 1, 2, and 3, were published from 72 Upper Circular Road, Calcutta, in 1886 and 1887.

17 *Imam Hasan and Imam Husyan* (Jibani), Ramanath Majumdar Street, Kalikata, 1901.

18 In my opinion, the literary oeuvre of Girishchandra motivated Bengal Muslims to take up the responsibility of writing sacred biographies in Bangla.

19 Shaikh Abdur Rahim, "*Bangabhasha O Muslim Samaj*", *Masik Mohammadi*, Bhadra 1336.

20 While much work has been done on the relevance of the ideology and functioning of a more 'secular' stint with modernity (Suchetana Bandyopadhyay 2010; Zahir Abbas 2011; A.K. Chakraborty, Neilesh Bose 2014), the relevance of religious intelligentsia in forming the dynamics of Muslim identity and modernity has not been properly discussed.

21 Abdur Rahim wrote *Islamtattwa* (*Islamic Theology*, 1896), *Namaztattwa* (*The Rules of Namaz*, 1898), *Hajj Bidhi* (*The Prescription of Hajj*, 1903), *Namajshikkha* (*Teachings of Namaz*, 1917), *Quran Hadiser Upadeshbali* (*Preaching of Quran-Hadis*, 1926), and *Rojatattwa* (*The Codes of Roja*, 1926).

22 The reception of his literary oeuvre in Bengal demands a separate domain of study, which should include biography of the Prophet's companions,

Islamic theologians, Persian poets, the history of Persian poetry, the time of Aurangzeb, and a travelogue on his trips to Turkey, Egypt, and Syria.

23 The translation and adaptation of *Fatwa-i Alamgiri* (first by Munshi Naimuddin) marks an attempt to standardize the Hanafi community with the juridical version of the scriptures. It also marks the disavowal of the syncretistic religious ideals of Akber and prioritization of a more rigid version of Islam that Auragzeb stands for. Aurangzeb assigned the task of compilation of *fiqh* to a collective of *Hanafi* jurists, the result of which was considered the source text for the reformation of the Hanafi community in the beginning of the twentieth century.

24 Though Dr Saiyyid Sajjad Husain (in Abu Jafar 1982: 132), the editor of *Comrade*, affirmed that Akram Khan collected his propositions in *Mostafa-Charit* from several English texts, in reality Akram Khan refuted the propositions offered by the books in English. Citing fifteen authors from the seventeenth century and twenty-seven more from eighteenth- and nineteenth-century England, Akram Khan affirmed that he wrote against them (Khan 2010: 88–95).

25 We must not forget that that Muhammad Akram Khan founded a periodical entitled *Ahl-e-Hadis* with an intention to disseminate *Mohammadiyah* ideals. There was massive antagonism between *Hanafi* (a periodical with a Hanafi worldview) and *Ahl-e-Hadis* over doctrinal issues and position vis-à-vis the British. *Hanafi* had a pro-British stance, while *Ahl-e-Hadis* was against the British.

References

Bangla sources

Abdul Jabbar, Shaikh. *Medina Sharifer Itihas*. Mymensingh, 1907.

Abdul Jabbar, Shaikh. *Islam Chitra O Samaj-Chitra*. Mymensingh, 1913.

Abdul Qadir. *Muslim Bangla Samayik Patra*. Dacca: Pakistan Publication, 1966.

Abdur Rahim, Shaikh. *Hazrat Muhammader Jiban O Dharmaniti*. Calcutta, 1887.

Abdur Rahim, Shaikh. *Eslam Tattwa*. Calcutta, 1889.

Abdur Rahim, Shaikh. *Islam Itibritta*, volumes 1 and 2. Calcutta, 1910, 1911.

Abu Jafar, compiler and editor. *Maulana Akram Khan*. Dhaka: Islamic Foundation of Bangladesh, 1982.

Ahmed, Muhammad Reyazuddin. "Mahasmashan Kabyer Bhumikay Islamer Abomanona". In *Bangiya Musalman Sahitya Patrika*. Bhadra, 1930.

Ahmed, Waqil. *Unish Shatake Bangali Musalmaner Chinta O Chetanar Dhara*. Dhaka: Novel Publishing House, 1983.

Ali, Abbas. *Kalir Fakirer Khela O Alemganer Nasihat*. Calcutta, 1920.

Anisuzzaman, editor. "Tritiya Bangiya Mussalman Sahitya Sammilan – Sabhapatir Abhibhasan" [The presidential lecture of Third Bengali Muslim Literary Convention], First Year, 4th issue, 1325 BS/1918. In *Muslim Banger Samayik Patra: 1831–1930*. Dhaka: Bangla Academy, 1969.

Abdullah al-Azad. *Mohammadi* (Falgun-Poush-Magh, 1327), Volumes 1, 2, no. II–III–IV. 1920.

Aziz, Abdul Maulavi. *Maydanal Ulum*. Calcutta, 1891.

Daad Ali. *Asheq-e-Rasul*. Calcutta, 1908.

Emdad Ali, Saiyyad. "Mahasmashan Kabye Anoislamic O Ashil Bhab". In *Bangiya Musalman Sahitya Patrika*. Baishakh, 1326BS, 1917.

Girishchandra Sen. *Mahapurush Muhammader Jiban Charit*, Volumes I, II, III. Calcutta, 1886–7.

Girishchandra Sen. *Imam Hasan and Imam Husyan (Jibani)*. Kalikata, 1901.

Hajarat Fatema Zohra (Ra) er Jiban Charit. Calcutta, 1928.

Haq, Muhammad Enamul. *Muhammad Enamul Haq Rachanabali*, Volume 4. Dhaka: Bangla Academy, 1995.

Islamabadi, Maniruzzaman. "*Echhlam O Shason- Adhikar*" [Islam and the rights over polity], *Mohammadi*, Year I, no. 3 (Poush), 1927.

Jabbar, Shaikh Abdul. *Medina Sharifer Itihas*. Gafargaon, Mymensing, 1314/1907.

Khan, Muhammad Akram. *Mochhlem Banger Samajik Itihas*. Dacca: Azad Publications Limited. 1965.

Khan, Mohammad Akram. *Mostafa-Charit*. Dhaka: Kakali Prakashani, 2010.

Munsuruddin, Muhammad. *Bangla Sahitye Muslim Sadhana*. 3 Volumes. Dhaka, 1960–6.

Murshid, Gholam. *Kalantare Bangla Gadya*. Kolkata: Ananda Publishers, 1992.

Rahim, Shaikh Abdur. *Hazrat Muhammader Jiban O Dharmaniti*. College Street, 1887.

Rahim, Shaikh Abdur. *History of the Moslem World: Islam Itibritto*. Calcutta, 1910.

Rahim, Shaikh Abdur. "*Bangabhasha O Muslim Samaj*", *Masik Mohammadi*. Bhadra, 1336/1929.

Reyazuddin Ahmed. *Hajrat Muhammad Mostafa*. Calcutta, 1924.

Sameeruddin, Munshi. *Bedaral Gafilin*. Calcutta, 1887.

Sen, Sukumar. *Bangla Sahityer Itihas*, Volume 4. Kolkata: Istham Publishers, 1965.

Sen, Sukumar. *Islami Bangla Sahitya*. Kolkata: Ananda Publishers, 1993.

Bangla periodicals:

Sudhakar

Mohammadi

Mihir

Bangiya Musalman Sahitya Patrika

English sources

Abbas, Zahir. *Construction of Bengal Muslim Identity in Colonial Bengal, C 1870–1920*. Bibliobazar, 2011.

Ahmad, Aziz. *Islamic Modernism in India and Pakistan, 1857–1964*. London: Oxford University Press, 1967.

Ahmad, Aziz. *Studies in Islamic Culture in the Indian Environment.* Oxford: Oxford University Press, 1969.

Ahmed, Rafiuddin. *The Bengal Muslims 1871–1905: A Quest for Identity.* Delhi: Oxford University Press, 1981.

Asad, Talal. *The Idea of an Anthropology of Islam.* Washington, DC: Georgetown University, 1986.

Balijon, J.M.S. *The Reform and Religious Ideas of Sir Sayyid Ahmad Khan.* Leiden: Brill, 1948.

Bandyopadhyay, Sekhar, editor. *Bengal: Rethinking History: Essays in Historiography.* New Delhi: Manohar, 2001.

Bandyopadhyay, Suchetana. *An Early Communist: Muzaffar Ahmed in Calcutta, 1913–1929.* New Delhi: Tulika Books, 2010.

Bayly, C.A. *The Origins of Nationality in Modern South Asia.* New Delhi: Oxford University Press, 1998.

Bray, Julia. *Writing and Representation in Medieval Islam: Muslim Horizons.* London and New York: Routledge, 2006.

Dey, Amit. *The Image of the Prophet in Bengali Muslim Piety 1850–1947.* Kolkata: Readers' Service, 2006.

Eaton, Richard. *The Rise of Islam and the Bengal Frontier, 1204–1760.* Berkeley: University of California Press, 1993.

Harlan O. Pearson. *Islamic Reform and Revival in Nineteenth-Century India: The Tariqa-i-Muhammadiyah.* Delhi: Yoda Press, 2008.

Hasan, Qamar. *Muslims in India: Attitudes, Adjustments and Reactions.* New Delhi: Northern Book Centre, 1987.

Jalal, Ayesha. "Identity Crisis: Rethinking the Politics of Community and Region in South Asia", *Harvard International Review* xxi, no. 3 (Summer 1999).

Keddie, Nikki R. *An Islamic Response to Imperialism: Political and Religious Writings of Sayyid Jamal Ad-Din "al-Afghani".* Berkeley: University of California Press, 1968.

Maitra, Jayanti. *Muslim Politics in Bengal: 1855–1906.* Calcutta: K.P. Bagchi, 1984.

Metcalf, Barbara Dally. *Islamic Revival in British India: Deoband, 1860–1900.* Princeton, NJ: Princeton University Press, 1982.

Mitra, Subrata K., Enskat, Mike, and Spieß, Lemens, editors. *Political Parties of South Asia.* Westminster and London: Praeger, 2004.

Rahim, Shaikh Abdur. *History of the Moslem World: Islam Itibritto.* Calcutta, 1910.

Robinson, Francis. *Islam and Muslim History in South Asia.* New Delhi: Oxford University Press, 1993.

Robinson, Francis. "Other-worldly and This-worldly Islam and Islam Revival", *Journal of the Asiatic Society,* Third Series 14, no. 1 (April 2004).

INDEX

Printed in the United States
By Bookmasters